WALK TO FREEDOM
Kriegsgefangenen #6410: Prisoner of War

WALK TO FREEDOM

Kriegsgefangenen #6410: Prisoner of War

JOHN L. LENBURG

Edited by

Jeff Lenburg

M☾☾NWATER PRESS

Copyright © 2002 John L. Lenburg
All rights reserved

Originally published as *Kriegsgefangenen #6410: Prisoner of War* by iUniverse
New edition published 2017 by Moonwater Press

Except as permitted under the U.S. Copyright Act of 1976, no part of this book may be used, sold, reproduced, distributed, or transmitted in any form or by any means, graphic, electronic, or mechanical, including abridgement, photocopying, serialization, recording, taping, dramatic, motion picture and other performing arts, or by any information storage retrieval system, including software and database, optical disk and videotext, or personal and commercial web sites, in any language, without prior written permission of the publisher.

Cover photo: The B-24 on fire is used for illustration purposes only. The actual event depicted is of B-24 Liberator from the 460th Bomb Group, 763rd squadron, shot down over Vienna on June 16, 1944.

Library of Congress Control Number: 2015930660

Publisher's Cataloging-in-Publication Data

Lenburg, John L.
  Walk to freedom: Kriegsgefangenen #6410: prisoner of war / John L. Lenburg. – 3rd rev. and expanded edition.
    pages cm
  ISBN: 978-0-9963206-3-4 (hardcover)
  ISBN: 978-0-9903287-4-2 (pbk.)
  ISBN: 978-0-9903287-5-9 (e-book)
  1. Lenburg, John L. 2. Bomber pilots—United States—Biography. 3. World War, 1939-1945—Prisoners and prisons, German. 4. Prisoners of war—United States—Biography. 5. World War, 1939-1945—Personal narratives, American. I. Title.
D805.G3 L369 2015
940.54`72—dc23
                                                  2015930660

Moonwater Press
P.O. Box 2061
Litchfield Park, AZ 85340
www.moonwaterpress.com

Printed in the United States of America.

I dedicate this book to the memory of three members of crew, Martin Troy, Ralph Wheeler and Rube Waits, and the fourteen crewmen of the three other planes who lost their lives in an air battle over southern Hungary on June 30, 1944. Also, to the many POWs who did not survive the "Black March of Stalag Luft IV" in the winter of 1945.

–John L. Lenburg

Inscription at a World War military cemetery in Europe reads…
"When you go home,
tell them of us and say
for your tomorrow,
we gave our today."

CONTENTS

Preface	xiii
Acknowledgments	xix

Part I: At War

1:	-1942- The U.S. Army Air Corps	1
2:	-1943- Basic Training and School	9
3:	460th Bomb Group	25
4:	-1944- Miss Fortune a B-24H	33
5:	Going Overseas	43
6:	Arriving at Spinazzola	51
7:	Flying Missions	59
8:	Blechhammer	93
9:	Going Down	99
10:	POW	111
11:	Royal Hungarian Hospital No. 11	117
12:	Sent to Prison	125
13:	Interrogation	133
14:	Riding 40 & 8 Style	137
15:	Stalag Luft IV	143
16:	Kriegie #6410	149
17:	Life in Kriegieland	153
18:	Christmas	165

19: -1945-The Long Walk	171
20: Freedom	187
21: Going Home	195
22: Epilogue	199

Part II: Crew Members

23: Alan Barrowcliff	211
24: Leonard Bernhardt	219
25: White's Crew	241
26: Evans's Crew	247
27: Champlin's Crew	251
28: Sorgenfrei's Crew	255
29: Jack Nagle	269
30: Capt. Leslie Caplan	273

Part III: Story of a Survivor

31: Fred Meisel	291

Part IV: Back to the Future

32: Fifty Years Later	297
Afterword	309
Photo Gallery	311

PREFACE

Oliver Wendell Holmes once said, "There is no time like the old time, when you and I were young."

In a way, this describes my days during World War II. I became engaged in a task that was not of my making, but one that I felt a duty to perform to the best of my ability. However, for this there was a price to pay—more for some than others.

I will always remember those days, some with pleasure, but too many with pain and tears. Some days I, too, remember more than others.

I wrote this book about what happened to me during our struggle against the oppressive Nazi regime of Germany for Allied air supremacy over Europe during the war. By writing this, I have tried to give an account of events that happened to others and me while serving in the U.S. Army Air Corps at that time. All of the material in this book is from notes and documents that I had saved or given to me over the past fifty years.

World War II was the biggest struggle against tyranny in the 7,000 years of the world's recorded history. It involved seventy nations of the world and more than 100 million people. One out of every twenty human beings on earth was involved in the war with over 10 million people killed. This Great War, fought on land, on the sea, and in the air, had produced the greatest armies, navies, and air forces that the world had ever known and most likely will never see again.

The European battlefields for Allied airmen during the war were high in the sky, deep in enemy territory. The air war with Germany was the most continuous part of World War II. From the battle of Britain on, it rose to epic proportions and remained so until the war's end.

With America's entry into the conflict, the war took on even greater proportions with 'round-the-clock aerial bombardment. The 8th and 15th bombed Germany by day and the Royal Air Force (RAF) Bomber Command by night, even though a terrible price was paid. More than 150,000 Allied airmen lost their lives with fifty percent of those who bailed out and survived wounded. Approximately 45,000 became prisoners of war of the Third Reich.

Every flyer that fell captive into enemy hands experienced the ordeal of escaping from an aircraft mortally damaged, about to go out of control or about to explode. Most of them only had seconds in which to react in order to save their lives. Some aircraft exploded instantly in balls of fire when hit and others would roll over and start a long downward spiral motion trapping those inside and sending them to their fiery deaths. Then some were lucky and the pilot was able to keep the badly damaged aircraft on a fairly straight and even course, so that the surviving crew could bail out.

The capture of an airman was uniquely perilous. He was usually alone, frequently dazed and injured, unlike captured soldiers and sailors captured by enemy soldiers and sailors who had shared the same life-threatening experience. The hostile action set upon him by the local population had been inflamed by the provocative rhetoric of their leaders. Any American airman, who became a prisoner of war of the "Third Reich" after March of 1944 until the end of the war, had to endure some of the worst deprivations of their lives.

While Germany officially signed the Geneva Conventions, they circumvented it with inflammatory speeches and newspaper editorials. In early 1944, Hitler and his propaganda minister, Dr. Paul Joseph Goebbels, branded all Allied flyers as "Luftgangsters" and "Terror Fliegers." He established a policy to murder captured airmen.

Goebells started publishing front-page editorials, charging that

Anglo-American attacks over Germany were no longer warfare but murder, pure and simple. He once said, "That it seems to us hardly possible and tolerable to use German police and soldiers against the German people when it treats murders of children with such leniency. Fighter and bomber pilots who are shot down are not to be protected against the fury of the people. I expect all police officers to refuse to lend their protection for these gangster type individuals. Authorities acting against, in contradiction to the popular sentiment, will have to account for their actions to me."

Many of the surviving airmen died in their bombers or fighter planes in battle. On the other hand, the many captured died at the hands of civilians or the SS, who hung, shot, or beat them to death. The actual number of murdered airmen is impossible to gauge but it was a significant number. In Munich, they began to hang captured Allied airmen. The bulk of war crimes perpetrated against military prisoners were against airmen. After Germany's surrender, the rush of the Cold War blunted the prosecution of many of these crimes. The need for justice never fully met.

There were primarily four POW camps. Stalag Luft I, Stalag Luft III, Stalag Luft IV, and Stalag Luft VI were for Allied airmen held captive in Germany. Luft I and III contained mostly Allied officers while Lufts IV and VI the enlisted men.

In his address to Congress, after the Japanese sneak attack on Pearl Harbor, President Franklin D. Roosevelt called the attack of December 7, 1941, "A day that would live in infamy." Well, a day that would live in infamy for me and some fifty other men of the 460th Bomb Group was June 30, 1944.

In the spring and summer of 1944, the air war over Europe was heating up. On June 30, 1944, a force of 450 heavy bombers and 150 fighter escorts left southern Italy to attack German oil refineries in Silesia in Poland. This was to be my thirty-sixth combat bombing mission. Unfortunately, we ran into a greater obstacle that no one anticipated: fog. Dense fog enveloped southern Hungary and, with visibility zero at 0945 hours, our bomber force was ordered to turn back.

On emerging from the fog, our formation unexpectedly came under fire by thirty-five to forty planes of the Royal Hungarian Fighter Group of the German Luftwaffe, over the Lake Balaton area of southern Hungary. They attacked our box formation by flying at us, four abreast, firing 20-millimeter cannons. Piloting the five aircraft in our box were Capt. John H. White, Jr., Lt. Elder Erfeldt, Lt. Robert G. Evans from the 760th Squadron, and Lt. Nelson H. Champlin and Lt. Ken Sorgenfrei of 762nd Squadron.

Because of this attack, four out of the five aircraft ended up damaged or lost, with seventeen men killed and twenty-four captured and becoming POWs. Of those killed, one died after his parachute would not open. Another captured airman Hungarian peasants savagely beat to death, with many who survived badly burned or wounded. Overall, we shot down or damaged six enemy fighters and killed one enemy pilot, Zsiros Gyula. Lieutenant Sorgenfrei, who flew the fifth plane, managed to escape this onslaught of enemy fighters. Altogether, the 15th Air Force lost seven aircraft that day.

This single event marked the beginning of a heart-wrenching time in my life that I would not soon forget: My life as a POW (or, Kriegsgenfangenen #6410, the code name given to me by the Germans).

When people find out that I was a prisoner of war in World War II, the most asked question is "What was it like?" By writing this description of events, I thought that it might help answer that question. I have also included material and statements from the crews involved in this ill-fated bombing mission to Blechhammer and testimony by Capt. Leslie Caplan, a captured American flight surgeon, held at Stalag Luft IV. Part I covers my life in the Army Air Corps, my POW experience and my liberation after the war ended. My book would not be complete without the remembrances and accounts of my fellow crew members. So Part II I have devoted to them, their memories, their recollections and statements about the war, namely: Alan Barrowcliff, Leonard Bernhardt, John H. White, Jr., Robert G. Evans, and Ken Sorgenfrei and their crews, plus Jack Nagle and Capt. Leslie Caplan, featuring testimony given by him to the War Crimes Commission about treatment and conditions at Stalag Luft IV. Part II is the testimony given by Captain Caplan to the War Crimes

Commission about treatment and conditions at Stalag Luft IV. Part III is a short biographical sketch of Fred Meisel. This man I met while in the Royal Hungarian Hospital No. 11 in Budapest. I think you will find his story interesting. The last part, Part IV, covers after the war, fifty years later.

Coningsby Darsen, a British soldier in WWI, once wrote in his diary, "To be forgotten - that is what I most dread. Never to have happened would not matter, but to have happened, to have walked the world, laughed, loved, created, to have taken part in events of history and then to be treated as though I never lived, there lies the sting of death."

John L. Lenburg
Portage, Indiana

ACKNOWLEDGMENTS

Many thanks to my wife Catherine Lenburg for her loving support, to my sons John, Jeff and Greg for their support and to my fellow crew members Alan Barrowcliff, Sparky Bohnstedt, Mike Brown, Jerry Conlon, Marvin Wycoff, Nándor Mohos and Bob Ingraham for their contributions to this project.

I also wish to credit the following sources that I used to research my book: my notes; the 460th newsletter, *Black Panther*, the *Ex-POW Bulletin* magazine, *The Shoe Leather Express*, *American Heritage: World War II*, the *Complete History of World War II*; the Indiana Historical Society, National Archives and Records Administration, and USAF Academy Library System; and declassified army documents that I obtained for writing this autobiographical account.

PART I:
AT WAR

-1942-
THE U.S. ARMY AIR CORPS

The year was 1942. A senior in high school, I had turned eighteen that year. America was at war after Japan's December 7, 1941 sneak attack on Pearl Harbor. Our country had suffered great military and material losses at the time. The country's mood was somber and things did not look good. The United States was moving towards a total war effort with the drafting of men for the armed forces, production of war equipment, rationing and civilian defense.

My Dad, Leo, and Uncle John, my Dad's brother, had joined the Civilian Defense Corps. A World War I veteran, Dad was a member of the American Legion Post 214 in Gary and was very involved in its drum and bugle corps. Consequently, I would tag along with him as he participated in many of the patriotic holiday parades and other functions.

In June of 1942, my best buddy Noble Allison graduated from Lew

Left: As a young boy growing up in Indiana. *Right:* Me with my sister, Joan, and my dad, Leo, in front of our 1927 Hupmobile. *Courtesy: John L. Lenburg Collection*

Wallace High School, which first opened its doors in 1926, in Gary, Indiana. Being a mid-term student, I still had another six months to go before graduating. Both of us had been in the Junior Reserve Officers' Training Corps (ROTC) infantry program at high school. It was also the year that I attended the Sixth Annual Hoosier Boys State at Indianapolis. I had been one of the students chosen by the Post 214. About five hundred boys from all over the state of Indiana attended this weeklong session. We set up a mythical state with two political parties, held caucuses and elected our own people to different governmental offices.

My interest in airplanes and flying started at an early age. Many a Sunday afternoon my sister Joan and I would pile into our 1927 Hupmobile and Dad would drive us to one of the local airports. There we would spend the afternoon watching planes take off and land. In those days, the airplane was still a new thing and people were

Watching the planes take off and land at Midway Airport in Chicago. *Courtesy: John L. Lenburg Collection*

My first model airplane at age 12.
Courtesy: John L. Lenburg Collection

fascinated at watching them take off, fly, and land. Besides, the country was still in the midst of the Great Depression and it was cheap entertainment on a Sunday afternoon. It was a real treat when my Dad would take us to Midway Airport in Chicago where we would watch the big planes fly. Certainly, those Sunday afternoon trips to the airport had a profound impact on me. I never stopped loving them. When I was 12 years old, I built my first model airplane. It was a balsa wood model. Years later, I would build hundreds of model planes and boats as a hobby.

Upon returning to school in September, my friend and classmate Harold Lane asked me to join him as a member of the flag squad at school. Every morning at the beginning of class, the school bell would ring. At this time, all of the students would rise and face the direction of the school flagpole. Dressed in our ROTC uniforms, Harold and I would hoist the American flag up the flagpole. The end of the school day, we followed same procedure when we took the flag down.

It was December 5, 1942; almost a year after America became involved in

At Fred and Irene Pope's farm in Chesterton, Indiana, 1939.
Courtesy: John L. Lenburg Collection

- 1942 - THE U.S. ARMY AIR CORPS

With classmate Harold Lane (center), dressed in our ROTC uniforms, taking down the flag at Lew Wallace High School as "the flag squad." *Courtesy: John L. Lenburg Collection*

the Second World War, I enlisted in the United States Army Air Corps to serve my country. My buddy Noble Allison and I, plus a group of my high school classmates, Andrew Anzanos, Robert Furry, Rudy Kurpis, Harold Lane, and Bill Lothian of the "Class of '43," Leo Joint, Jr. and Bruce Reibly of the "Class of 42," enlisted in the U.S. Army Air Corps. We were part of a contingent of sixty-two men and the last group of enlistees to leave Northwest Indiana for the service. My classmates and I wanted to go into the U.S. Army Air Corps, so we had decided to enlist. By enlisting, we could pick the branch of the service that we wanted to go into. Starting in January 1943 the armed services would rely on the draft only for their manpower needs. Out of our group, Harold Lane, who flew B-17s with the 8th Air Force out of England, was the only one of our group killed in action of which I am aware.

I had only a few more months of school left before finishing in February and knew would be drafted because of the war anyway.

So we talked it over with our school principal, Miss Verna Hoke, to make sure that we would get our diplomas in June before taking this step. She assured us that we would, even though, technically, we would be finished in February. Little did I know at the time, when June rolled around, I would be graduating from an aircraft mechanics school in Los Angeles.

Six days later, we loaded on a Pennsylvania Greyhound Lines bus by the post office building in downtown Gary for our trip to the Indianapolis induction center. At the time I enlisted, my father was in the hospital. If I had known how seriously ill he was then,

Local newspaper clipping touting our departure as the last group of enlistees to leave Northwest Indiana. *Courtesy: John L. Lenburg Collection*

Left: Postcard view of Company Street at Fort Benjamin Harrison; *Right:* Graduates of the U.S. Army Chaplain School at Fort Benjamin Harrison pose for a group picture, April 1942. *Courtesy: Library of Congress Prints and Photographs Division*

I probably would not have left at that time as a new enlistee. After spending the first night at the Barnes Hotel in downtown Indianapolis, they took us to the induction center at Fort Benjamin Harrison the next morning. There on December 12, 1942, they swore us in and gave us physicals and shots.

It was not long before I was on kitchen patrol (KP) and guard duty. That is when my real army experiences began. On my first stint on guard duty, the outside temperature was well below zero. You were on guard duty for twenty-four hours, which amounted to two hours on, and then four hours off. For KP duty, you were on for the entire day.

At Fort Harrison, where they stationed me, they interviewed me and gave me a battery of skill and aptitude tests. I requested they put me into a flying unit and soon they granted my request.

With the class AM-16 at the Aero I.T.I. School in Los Angeles in 1943; I'm sitting on the end, right hand side. *Courtesy: John L. Lenburg Collection*

-1943-
BASIC TRAINING AND SCHOOL

Our group shipped out for basic training in Miami Beach, Florida after spending about a week at the Fort Harrison induction center. During our stay in Miami Beach, they housed us in the Copley Plaza Hotel on Collins Avenue.

All we seemed to do in basic training was march, have more shots, march and do more guard duty. We practiced marching on the golf courses in Miami Beach. All of us would sing in cadence while marching. The songs we sang included "The Army Air Corps Song," "The Caissons Go Rolling Along," "Over There" and "Pack Up Your Troubles."

We also were required to pull guard duty by patrolling the beach at night. We had to be on the lookout for German submarines that might be trying to land fifth columnists or saboteurs. One afternoon, after having been there about three weeks, they called me into the orderly room. They

Top: The Copley Plaza Hotel in Miami. *Bottom:* January 1943: Noble and me, Miami Beach. *Courtesy: John L. Lenburg Collection*

quickly informed me that my father had passed away and that they were giving me an emergency furlough, so I could attend his funeral. The news of my dad's sudden death hit me like a ton of bricks and I broke down.

Enjoying time on the beaches of Miami. *Courtesy: John L. Lenburg Collection*

Upon returning to Miami after attending my father's funeral, I found that all of my schoolmates had shipped out. After completing my basic training, I shipped out with another group of G.I.'s on a troop train to Los Angeles, California. It took us five days to get there, after which they sent us to an aircraft mechanics school, Aero Industries Technical Institute (I.T.I.), a division of Aero-Crafts Corp., in Los Angeles. It was a privately-run school under contract with the Army Air Corps to teach us aeronautics. I was at the school from February 8 through May 23, 1943. We took courses in sheet metal, aircraft structures, electrical systems, hydraulics, engines, and instruments.

Normal class size at the Aero I.T.I. school was eighteen men but somehow, we only had nine. This was called a "SNAFU" meaning "Situation Normal All F*? %#D Up." I was in class AM 16. Going to school here was a good deal since we did not have to pull guard duty, KP, or do our laundry. We also had what they called a class "A" pass, which meant we were free to leave the school in the evenings and on weekends.

This gave me a chance to hitchhike all over Southern California. The opportunity to go to school in Los Angeles offered many interesting experiences, especially with all of the glitter of Hollywood nearby.

Top: An aerial shot of the Aero I.T.I. school complex. *Bottom:* The barracks where I stayed at Aero I.T.I. in Los Angeles. *Courtesy: John L. Lenburg Collection*

An original 1941 trade ad for Aero Industries Technical Institute where U.S. Army Air Force enlistees trained.

Front view of one the hangars at Aero Industries Technical Institute.
Courtesy: U.S. National Archives and Records Administration

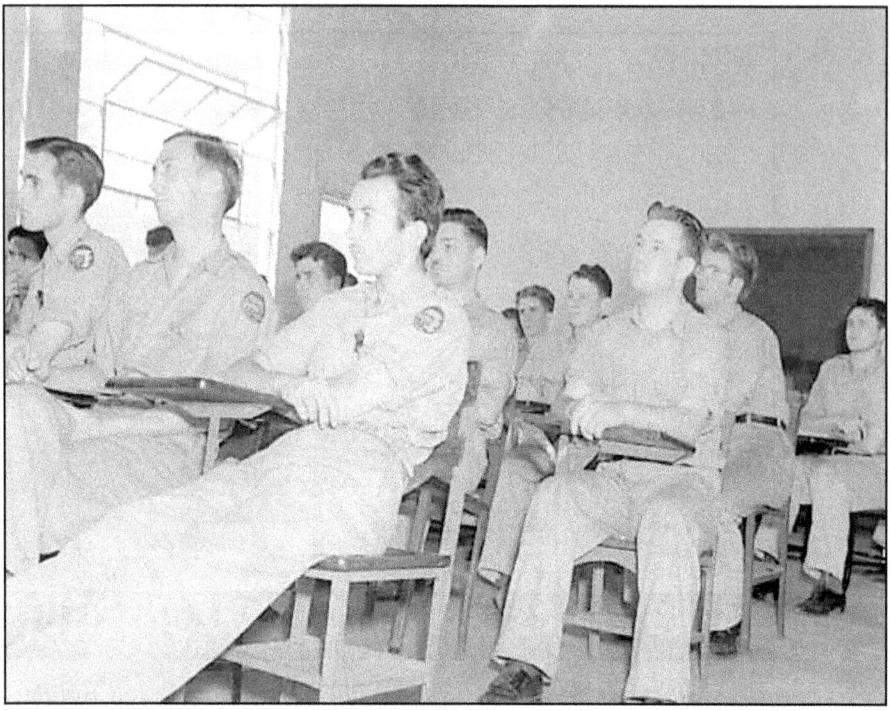

Top: Training included courses in aeronautics. *Bottom:* Aero I.T.I.'s busiest shop was its riveting and sheet metal shop to train riveters for the aircraft factories to build planes to aid the war effort. *Courtesy: U.S. National Archives and Records Administration*

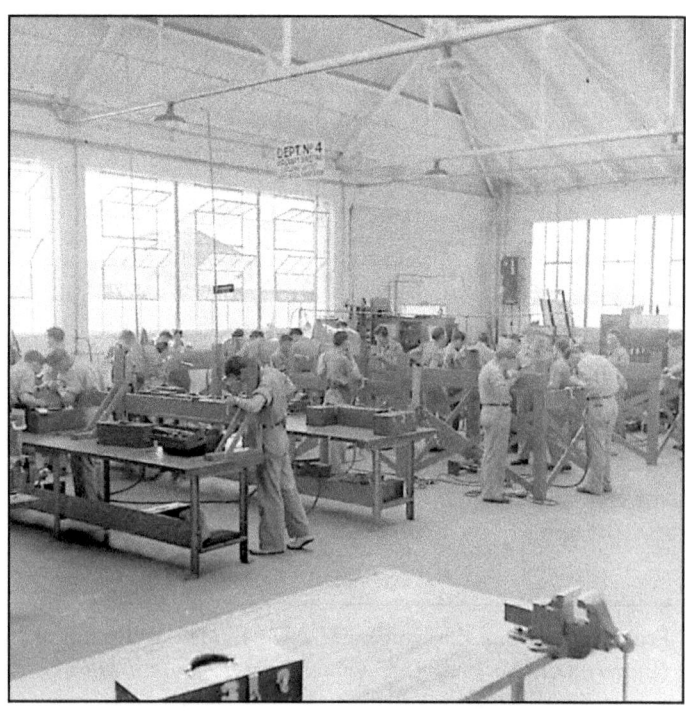

Original poster for the Republic Pictures wartime film, *Thumbs Up* (1943).

I did manage to spend quite a bit of my free time there. One of my experiences was spending a day at Republic Pictures movie studio. Republic was shooting a wartime musical called, *Thumbs Up*, a story about the Royal Air Force (RAF), a service branch of the British Armed Forces, in the war. Released on July 5, 1943, the 67-minute musical drama starred Brenda Joyce, Richard Fraser, Elsa Lanchester and Arthur Margetson and Joseph Stanley directed the picture.

The studio contacted our school to see if they would loan one of our planes to them for this picture. School authorities agreed, so my class was picked to help with the project. One of our instructors who went with us happened to be a friend of the director's. We dismantled the plane by removing the wings, so they could transport it Republic Pictures studios. Once there, we would reassemble it. We spent the day at the studio until they were finished using the plane in the scenes they filmed.

April 1943, overlooking the famous Hollywood Bowl in Hollywood, California.
Courtesy: John L. Lenburg Collection

We met various movie stars during a tour of the studio, where we they treated us royally. One of those stars was Roy Rogers who was filming a western. Of course, this was a big thrill for a boy who had been born and raised in the Midwest and used to spend Saturday evenings at the Strand Theatre, in Hobart, watching cowboy movies.

While stationed there, I was able to visit the famous Hollywood Canteen, a club that offered food, dancing, and entertainment for servicemen, many times and attend live broadcasts of popular radio shows. One of those was *The Charlie McCarthy Show*. The fact I was a roller skating buff in high school, I had my skates shipped to me. This gave me something else to do since there were roller rinks in Hollywood, Glendale and Pasadena.

Graduation was the end of May and I graduated with a grade point average of 84 percent. I was "tops" in our class graduating with honors. I had been taught about all aspects of the airplane and promoted to PFC. Aero I.T.I. awarded me a scholarship to continue my studies, if I so desired, at the school after the war.

> **Graduates as Air Mechanic**
> Pvt. John L. Lenburg, 18-year-old Gary youth, has been graduated from Aero Tech in Los Angeles, according to the army air forces technical training command. He has since been assigned to duty with an active air forces unit.
> Specializing in aircraft engines, Private Lenburg received a special diploma signed by Maj. Gen. John F. Curry, commanding general of the 4th district, AAFTTC, in recognition of his outstanding record.
> In addition, he was given a special scholarship award by Ralph Hemphill, Aero Tech president, which will permit him to continue his study of aircraft mechanics after the war.

Usually, they would ship out right after graduating but due to some SNAFU, our class did not. Maybe they were waiting for the other half of our class to finish. We ended up staying a couple of extra weeks. When our classes shipping orders finally came through, they put us on a train and shipped us to the 309th Depot Repair Squadron at Norton Air Force Base in San Bernardino. After arriving at the train station in San Bernardino, they took us by truck to the air base. Upon our arrival, they informed us the 309th was no longer there. Was this another SNAFU? The outfit had shipped out several days earlier to March Field in Riverside, which was about fifteen miles away. So, they loaded us on a truck again and took us to March Field. The 309th was there but they were getting ready to leave the next morning for the Mojave Desert for ten days of maneuvers. When finished with the maneuvers, the outfit was to leave for Egypt. Things really went into high gear for me. I was issued my full complement of clothing and equipment and still be ready to leave the next morning.

Early the next morning, our outfit trucked out to the edge of the desert and made to hike to the area where we set up camp. At roll call the next morning, they informed us that there would be no promotions or transfers out. On the tenth day, after spending nine days there battling scorpions, rattlesnakes and flour bombs dropped by the air force, during roll call, my name was called and I was told to be ready to ship out. I was shipping out to the 499th Air Base Squadron Replacement Pool at the San Bernardino Air Base. So, I left all of my buddies behind and they transported me by jeep to the air base. (Another SNAFU maybe?)

Following a few weeks of lying around, I put in for gunnery school. Now I had put in for flying status at Fort Harrison, I went through a battery of tests, which came out fine. So far, nothing had come of it. Several weeks later after roll call, I decided to check with the orderly room to see if any shipping orders had come in for me.

The sergeant asked me my name and then checked over the shipping roster. "Didn't they call your name at roll call this morning?"

I said, "No."

"Well, you're supposed to catch the ten o'clock train for Las Vegas this morning. You go and pack and I'll get a driver to pick you up in a jeep and take you to transportation headquarters. They will give you a voucher for you to give to the ticket agent at the train station, so he can issue you a rail ticket." (I have a feeling of another SNAFU was in the making.)

After rushing around, I arrived at the train station and handed my travel voucher to the ticket agent. He looked at it, and then returned it to me and said, "I can't give you a ticket for this voucher. It is made out for wrong railroad."

It was a good thing my driver waited. Back to transportation department we flew. I explained to the clerk what was wrong so they issued me a new voucher. We rushed back to the station just as the train pulled in. It was off to Las Vegas to gunnery school.

July 1943: Wilted in Las Vegas after being stationed at Nellis Air Base outside of town. *Courtesy: John L. Lenburg Collection*

The train arrived late in the afternoon, so I walked up to the MP at the station telling him who I was. He looked over a roster sheet that he was holding. After flipping through several pages, he said, "What are you doing here? You're not due till Sunday evening and this is only Friday." (Definitely another SNAFU).

I asked him what I should do. He told me to show up Sunday evening.

Some Hobart friends of my parents, Adolph and Mary Blakeman, had moved with their son James and daughters Elizabeth and Maxine to Las Vegas in the early 1940s. I called and told them. They said that I should come and stay at their house for the weekend and so I did. While staying there they gave me a tour of the casinos and Boulder Dam (now Hoover Dam). Sunday evening, they took me back to the train station where I reported to the MP. He told me to jump on the truck and off I went to the air base.

Stationed from July 26 through September 6, 1943 at Nellis Air Base located outside of town in the desert, I went to Student Training Squadron 8 class 43-36 for training as an aerial gunner there. The training was intense during the first six weeks when they taught us how to operate thirty- and fifty-caliber machine guns, a gun turret, and aircraft recognition besides taking plenty of target practice.

The last two weeks we spent at the gunnery range in Indian Springs, Nevada, learning how to shoot at objects in the air by shooting aerial tow targets.

The highlight of my training was flying in B-17s. I was now flying every day and sometimes twice a day. My first flight in a plane went without a hitch; in fact, I was thrilled with it. I cannot say that for all of my classmates. One in particular was Denny Dimowitz. He was from Brooklyn, New York and was always bragging about how he had flown piggyback in P-38s. The P-38 was a twin-engine fighter plane. On some models, they had taken some of the radio equipment out and squeezed another seat in behind the pilot.

Well, all of us lined up at the flight line to wait our turn for our first flight and greet our comrades who were returning. I had already made my flight. When it became Denny's turn, he turned and gave us the old thumbs up sign and climbed into the cockpit. Each flight took about thirty minutes. About a half-hour later, the plane landed with Denny. The pilot taxied the plane and stopped in front of us. He broke out laughing and shaking his head. All we could see in the rear cockpit was a big pile of something that looked like a sheet.

After shutting off the engine, the pilot jumped out and walked up to us still laughing. Someone asked him where Denny was. The pilot said, "He is still in the rear cockpit under his parachute. He got so sick and excited that he urped all over the plane and pulled the ripcord on his parachute. It almost pulled him out."

Denny had to clean up the plane with a bucket of water. He never bragged about his flying again. In fact, he took a lot of ribbing.

Another funny incident happened one night about an hour after lights out. Flying seemed to be on everyone's mind almost constantly since it was new to all us. The fellow in the top bunk next to me rose up and perched himself on the edge of the bunk and rolled out headfirst. He hit the floor with a large thump, which caused everyone to jump up out of bed. He stood up and said, "The darn parachute didn't open." He was not hurt, except for a large bump on the top of his head. Of course, we all had a good laugh. Another not so funny incident that happened to me was the time my flying restraint belt broke. We were flying in two-seated training planes, an AT-6, with the pilot in the front cockpit and me in the rear and a thirty-caliber machine gun mounted on the side. They would fly us in a four-plane stepped down formation. The high plane would peel off and make a pass at a tow sleeve pulled by another plane. At this point, I was supposed to fire all my rounds of ammunition at the target. When finished, I would wiggle my gun up and down as the pilot would peel off and fly the plane back into formation.

Flying B-17s every day and sometimes twice a day was the highlight of my training at Nellis Air Force Base. *Courtesy: John L. Lenburg Collection*

During this time, I was to take out the empty ammunition can and replace it with a full one. In doing this, you had to stand up in the cockpit while almost flying upside down and finish in time for the pilot to make a second pass at the target. This whole process happened very fast. You wore a belt around your waist that had a belt attached to each side that snapped in a ring on each side of the cockpit. This held you in the plane while the pilot was doing all of this maneuvering while you changed the cans of ammunition.

My belt broke after we made our first pass. I could not communicate with the pilot about what happened. I hung on for dear life trying to change my ammunition and to be ready for our second pass at the target. One thrill I did not care to repeat.

I was glad to see September roll around, for the days and evenings started getting a little cooler. Stationed in the desert during the hot summer months was the worst time to be there. The air base had no air conditioning in those days.

Upon graduating, they awarded me my wings, promoted me to sergeant, and shipped me to the 18th Replacement Wing Squadron L at an air base near Salt Lake City. One morning, I was singled out and told, "Get ready to ship out." I would be leaving all of my gunnery classmates behind. I was excited for I knew that my next assignment would be at a B-17 base since I had trained for them. Wrong! When I arrived at the air base in Clovis, New Mexico, I found it to have only B-24s. Was this another SNAFU?

Clovis, New Mexico was just a large bump in the road in those days. There was nothing there, except the air base. The most exciting thing that they had for entertainment was a roller rink in Portales, New Mexico that I used several times.

On September 14, 1943, they assigned me to the 302nd Bomb Group, 356th Bomb Squadron, and, on November 5, 1943, I completed my first phase of training, during which I housed in a tent with five other men. You had to keep the sides of the tent rolled up during the day because there was nothing, but red sand and the wind blew all day long. They started training me to be a flight engineer/gunner on a B-24. I had

The next destination: Clovis Army Air Field for B-24 training. *Courtesy: United States Air Force*

to redo most of my training since a B-24 had different types of systems than a B-17 and they put me on a crash course. They had me either in a classroom or up in the air with an instructor. I soon found out what they had in store for me.

Erf's Crew

Flight Crew #3-8

1st Row Kneeling (L to R):

2nd Lt. Elder "Erf" A. Erfeldt**	0806051	Pilot
2nd Lt. Alan P. Barrowcliff	0808340	Co-Pilot
2nd Lt. Marshall J. Brown	0754876	Bombardier
2nd Lt. Matthew L. Hendricks**	0696732	Navigator (Not in picture)

2nd Row Standing (L to R):

T/Sgt Ralph F. Wheeler*	12034660	Radio Operator/Right Waist Gunner
S/Sgt John "Jack" M. Nagle Jr.	8187526	Asst. Engineer/Nose Turret Gunner
S/Sgt Rube Waits Jr*	139369	Asst. Radio Operator/Ball Turret Gunner
S/Sgt Martin Troy*	313507	Armament /Left Waist Gunner
Sgt Leonard "Pappy" Bernhardt*	31307804	Asst. Armament/Tail Turret Gunner
T/Sgt John L. Lenburg	15383056	Engineer/Top Turret Gunner

* killed in action

* * deceased

-1943-
460TH BOMB GROUP

After several weeks of intense training, I passed the test of becoming a flight engineer/gunner on a B-24 and earning a promotion to the rank of staff sergeant. On October 18, 1943, the Army Air Corps activated crew 4-0-93, to which they assigned me, under the command of a pilot named Lt. Elder "Erf" Erfeldt, also trained to fly B-24s. On November 9, they assigned us to the 460th Bomb Group, 760th Bomb Squadron, becoming the nucleus of a ten-man bomber crew #3-8.

At that time, we started making many days and night training flights. During one these flights, an incident happened that scared the heck out of me. On a moonless night, I was flying with a major as the pilot, Erf as co-pilot and myself as the flight engineer. We were making simulated bomb runs on a target in the desert. The major asked me to go to the rear of the plane and make an equipment check. Before leaving

Black Panther 460th Bomb Group insignia patch worn by crew members.

I told him not to open the bomb bay doors until I returned.

After completing the check, I opened the rear door that separated the bomb bay from the rear of the plane. I said to myself, "Gee, it's awful breezy," but proceeded walking across the catwalk to the front of the plane. There were no lights in this area of the plane. It was pitch black. As I was feeling my way along, I looked down and suddenly I noticed a light going by. It turned out that the major had opened the bomb bay doors. The lights were from some ranch house on the ground below. At that point, I felt something to grab on to, and slowly made my way back to the flight deck. I gave the major a piece of my mind since I was wearing no parachute.

Before completing our training, another member added to our crew was Lt. Alan Barrowcliff, our co-pilot. Upon completion of our training, we shipped out to Chatham Army Air Field, Savannah, Georgia, on a ten-day delay traveling there. I took the time to go home for a visit, hitchhiking to Amarillo and catching a commercial flight to Midway Airport in Chicago, before reporting.

Ten days later, I arrived at Chatham Field in Savannah, Georgia. Here, we picked up the rest of crew #3-8: Lt. Marshall (Mike) Brown, bombardier; Lt. Matthew Hendricks, navigator; S/Sgt. Ralph Wheeler, radio operator/waist gunner; his assistant Sgt. Rube Waits, ball turret gunner; S/Sgt. Martin Troy, armament\waist gunner; his assistant Sgt. Raymond Terry, tail turret gunner; and my assistant, Sgt. Jack Nagle, nose turret gunner. Jack, it turned out, was a relative of the famous Wright Brothers.

In the following weeks, I was better acquainted with the new crew members and began training by flying as a team. Terry was taken

Left: Barrowcliff and Hendricks. *Right:* Waits by the ball turret. *Courtesy: John L. Lenburg Collection*

Our crew chief Philip Baideme. *Courtesy: John L. Lenburg Collection.*

Top: 460th Bomb Group personnel congregate at the unit bulletin board. *Bottom:* 460th Bomb Group airmen train to load 100-pound bombs at Chatham Field in December 1943.

from our crew, for reasons that I do not remember. Pvt. Leonard Bernhardt replaced him. Thus, we would become known as "Erf's crew."

We practiced formation flying, gunnery, simulated bombing missions and navigational flights, day and night. Meanwhile, they promoted me to tech sergeant. My responsibility was to keep the flight log, make a visual pre-flight inspection of the aircraft before each flight, load the plane and its fuel, assist the pilot and co-pilot during takeoff and landings, monitor fuel consumption and keep check on the electrical and hydraulic systems while in flight. In combat, I was to operate a Martin top turret with its two fifty-caliber machine guns, located on the flight deck behind the pilot.

December 1943 was a busy month for me. I served as the best man at the wedding for our assistant flight engineer and colleague Jack Nagle.

Left: With Jack's dad. *Right:* With Jack and Jack's brother on my visit to Houston. *Courtesy: John L. Lenburg Collection*

December 1943: On our way to Havana, Cuba, for the flight we called, "The Rum Running Trip of '43." *Courtesy: John L. Lenburg Collection*

He got married on Christmas Eve in Savannah, Georgia. The 460th made a group formation flight to Havana, Cuba also this month. Erf told me that Colonel Crowder, our group commanding officer, was going to lead our group of planes and he was flying our plane. He was a West Point graduate and very strict.

We were all on pins and needles since he picked our crew to fly with him. I was to be his flight engineer; Erf his co-pilot. We hauled in our plane back to the States cases of Cuban Rum on our return from Havana, for the Colonel. We called this flight, "The Rum Running Trip of '43." On Christmas, since this would be our last Christmas in the States, the colonel distributed a bottle of rum to all of the crews.

We learned that the air echelon, part of the 460th, was going to

Mitchell Field on Long Island, New York after Christmas for overseas staging and processing and the ground echelon to Camp Patrick Henry. They embarked from Hampton, Virginia and traveled in a 100-ship convoy that encountered enemy subs during their crossing.

Rumor had it that the air echelon might be going to Italy, by way of South America and Africa. So, I wrote and alerted my best friend Noble, stationed at an air base in Natal, Brazil, who kept in touch.

MISS FORTUNE

Model - Consolidated B-24H Liberator
Built in Fort Worth, Texas.
Serial # - 41-29291
Group insignia Black Panther painted on each side of nose.
Call letter, "V" for Victor, painted on each side of the plane.
Tail markings - bottom half-yellow circle - top half painted yellow with an unpainted square in the center.
460th Bomb Group (H)
760th Bomb Squadron (H)
55th Bomb Wing (H)
15th Air Force
MTO

NOTE: The photo of "Miss Fortune" is a recreation of the actual aircraft.

-1944-
MISS FORTUNE: A B-24H

O n January 1, 1944, Erf and I flew to Hunter Municipal Airfield, Savannah, Georgia. There we picked up a new B-24H, serial #41-29291. The B-24 had been made at the Consolidated Vultee Aircraft Corp. plant in Fort Worth, Texas, and was nicknamed the "Liberator." It was a four-engine, heavy bomber powered by four Pratt and Whitney R-1830 radial engines and weighed 38,000 pounds. This monolith could carry a bomb load in excess of 12,000 pounds and, when fully loaded for a mission, would weigh close to 70,000 pounds.

Carrying a crew of ten people, the plane featured a top turret, nose turret, ball turret, and a tail turret. Each turret was fitted with two fifty-caliber machine guns with two of them mounted in each waist window for the waist gunners. In its day, the plane was fast for its size: It flew at 306 mph between 20,000 to 32,000 feet in altitude and carried approximately

Top: B-24 Liberators enter final assembly at the Consolidated Vultee Aircraft plant in Fort Worth, Texas. *Bottom:* (left) The company plant logo; (right) Propellers are installed outdoors on B-24s in production. *Courtesy: The University of North Texas Libraries*

2,343 gallons of gasoline, with two auxiliary tanks carrying an additional 225 gallons each. The wingspan was 110 feet, the length approximately 66 feet and range 2,850-plus miles. Throughout the Second World War, American and British airmen relentlessly flew their Consolidated B-24 Liberators into the skies over Europe and the South Pacific to attack the Axis strongholds. While overshadowed by the highly publicized exploits of Boeing's "Flying Fortresses," the incredible B-24 saw action over more operational fronts than any other bomber.

Interestingly, the inadequacies of the B-17 eventually spawned the Liberator. The Consolidated B-24 Liberator would become one of the most famous bombers of the Second World War. More than 18,400 B-24 bombers were built, making it the most produced American wartime aircraft of its time. The aircraft was built and assembled at five plants

The public sees the B-24 Liberator up close at the Consolidated Vultee plant in San Diego, California, in 1944. *Courtesy: The University of North Texas Libraries.*

December 29, 1939: The B-24 Liberator takes a maiden test flight at Lindbergh Field in San Diego, California. The flight lasted seventeen minutes. *Courtesy: Time-Life Pictures*

in the U.S., including Consolidated Vultee, Fort Worth, Texas; Consolidated Vultee, San Diego, California; Douglas Aircraft Co., Tulsa, Oklahoma, Ford Motor Co., Willow Run, Michigan; and North American Aviation, Dallas, Texas.

On December 29, 1939, the B-24 Liberator took its first test flight at Lindbergh Field in San Diego, California. The flight lasted seventeen minutes. The U.S. Army Air Corps took its first delivery of the B-24s in mid-1941 and shortly deployed them into service. This remarkable aircraft was neither sleek nor graceful. It relied on a magnificent wing design that not only improved the aircraft's operational range but reduced drag as well. The long, high aspect ratio Davis wing was not the only development that amazed the Army Air Corps, for this incredible bomber possessed tricycle landing gear, two slab-like rudders, and two cavernous bomb bays that ended all doubt that the B-24 was, indeed, a heavy bomber.

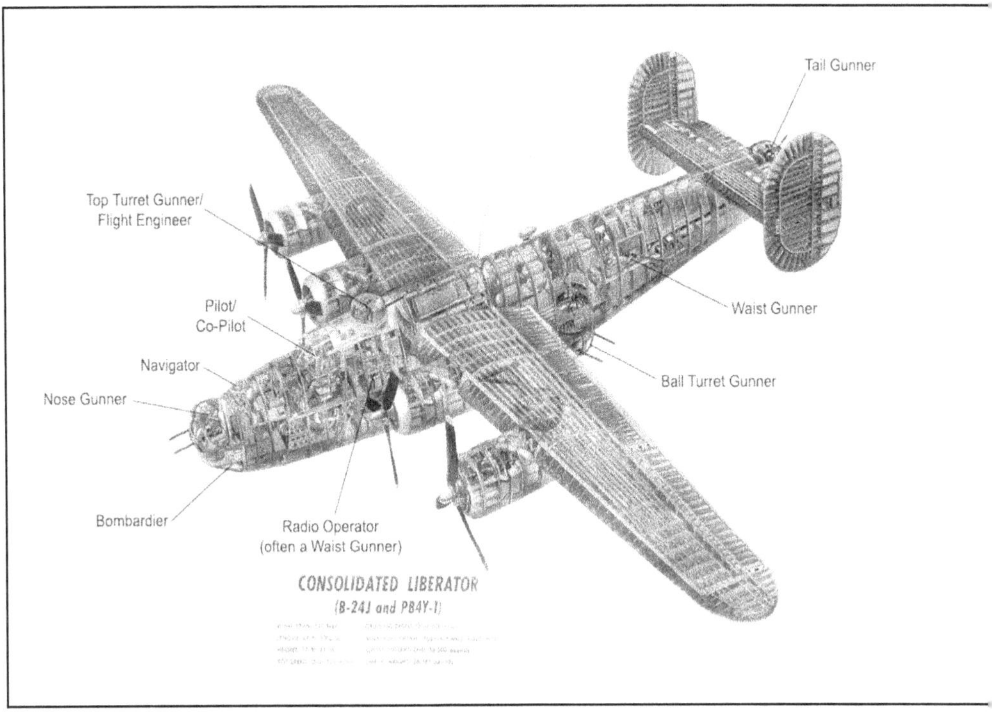

Top: The B-24J general layout and crew positions like that of "Miss Fortune." *Bottom:* View of the cockpit. *Courtesy: John L. Lenburg Collection*

View of the tail gunner turret of the B-24 Liberator. *Courtesy: U.S. National Archives and Records Administration*

During the service life of the aircraft, the deep, long fuselage enabled the U.S. Army Air Corps to adapt the Liberators to an endless variety of wartime tasks. They served admirably, not only as strategic bombers, but also as photoreconnaissance aircraft, anti-submarine patrol ships and cargo transports as well.

Though the initial versions of the B-24 deployed to the Mediterranean and Great Britain, the desperate need for long-range aircraft in the South Pacific caused the Liberator to become the mainstay of Allied operations in the island campaigns. By 1943, the B-24 had been replaced by the B-17 as the standard long-range heavy bomber in the South Pacific. Piloted by Army and Navy aircrews, as well as British aviators, the Liberators fought valiantly until replaced by the massive waves of silver B-29s.

As with most aircraft, the B-24J was a refinement of the previous Liberators, but during the production run of the "J" model, the five aircraft factories in existence constructed more than 6,600 aircraft of this version. Even though all B-24Js appeared similar externally, wartime priorities created variations in components and numerous differences in the aircraft produced. The long wing that enabled the B-24 to carry such

Ford factory workers assemble a B-24 at the Willow Run, Michigan plant, one of five aviation production facilities in the U.S. that built the Liberator during the Second World War. *Courtesy: Motor Trends*

An outside view of the tail gunner position of the aircraft. *Courtesy: U.S. National Archives and Records Administration*

massive bomb loads proved to be unable to absorb a great deal of battle damage, but the fuselage and tail assembly were unbelievably rugged.

The Liberator proved to be one aircraft created during the Second World War that was capable of performing such a diversity of missions. Although the end of the war sounded the death knell for the old warriors, the B-24 Liberators created a unique chapter in aviation history.

By March of 1944, the Ford Willow Run plant was reportedly producing B-24H Liberators at a rate of every 100 minutes, seven days a week, with the supply of B-24s that exceeded the U.S. Army Air Corps' ability to use them in the war. By mid-1944, the San Diego and Willow Run plants had the capacity to deliver more than enough B-24s, resulting in the closure of the Douglas at Tulsa and North American at Dallas lines. Fort Worth continued to produce B-24Js until the end of the 1944.

After arriving at Hunter Municipal Airfield in Georgia, Erf and I flew the new 18-ton B-24H Liberator back to Chatham Field and the flight took thirty-five minutes. We would eventually name it "Miss Fortune." One of the first things that Colonel Crowder did was have the group insignia of Black Panther painted on both sides of the nose of each plane. We spent several more weeks flying our new ship and getting acquainted with it. Philip Baideme was assigned as our crew chief. He was responsible for maintenance of the plane.

On several occasions, while making cross-country training flights to get better acquainted with our new ship, Erf would have me slide into the pilot's seat and fly the plane. He said that he wanted me to be able to fly our plane in case of an emergency.

"Miss Fortune" bears the Black Panther insignia painted on both sides of the nose of each B-24 in the 460th Bomber Group. Pictured: Jack Nagle in the nose turret of our plane.
Courtesy: John L. Lenburg Collection

He also taught me how to keep our plane on course by using the radio beacon. Once in combat, you never know when the flight engineer might need to assist in flying the plane back to the base. This happened on several occasions.

A couple of tough guys, Troy and me, in North Africa. *Courtesy: John L. Lenburg Collection*

-1944-
GOING OVERSEAS

Our crew represented a pretty good cross-section of the U.S. Erf was from Oregon; Barrowcliff, Delaware; Brown, California; Hendricks, Michigan: Nagle, Texas; Wheeler, New Jersey; Waits, Georgia; Troy, Connecticut; and Pappy, Massachusetts and me, Indiana.

With our training in the States over, our crew was beginning to jell as a team; everyone was working as a team member. Camaraderie and the morale of our crew were very good. Some members were the recipients of some kidding, but all took it in stride.

We had a good pilot in Erf and a lot of confidence in him as our team leader. Erf was not as militaristic as some other officers in our group. He did not require us to salute him, but we respected him. Hendricks, our navigator, was the only one who did not seem to fit in.

Erf tried to have him replaced but was not successful.

We arrived at Mitchell Field a few days after the first of the year. Luggage racks had been installed in our bomb bays to carry our clothing and personal things. New York would be the point of our embarkation for our trip overseas. We spent several weeks getting our shots, teeth checked, among other things.

While there, I spent a lot of my spare time in New York City. A briefing followed during which they showed us a map of our overseas route. The first stop would be Morrison Field, West Palm Beach, Florida, then Waller Field in Trinidad, Belém and Natal, Brazil, and Dakar, West Africa.

On January 22, 1944, with Erf at the controls and our plane loaded, we took off from Mitchell Field along with sixty-one other B-24s from our group. Installed in our bomb bays were baggage racks in order that we might carry all of our belongings plus some group equipment.

Besides our crew, we also carried four extra passengers: our crew chief Philip Baideme, John Edwards, Franklin Lichtenhan and Sam Goldhagen.

Three days later, on January 25, we left Morrison Field for South America. Erf was given sealed orders before our departure to be opened an hour after leaving Morrison Field. I wondered as I saw the ground disappear what lay ahead for me. An hour out of Morrison Field, Erf opened up our sealed orders. It gave us our destination of Oudna Air Base in Tunisia with a stop in Marrakesh, Morocco (also known as Marrakech). They assigned us to the 15th Air Force with our final destination being southern Italy.

The 15th Air Force activated at the conclusion of the North African campaign, thus flying its first missions from North African bases. Using North Africa as a base the Allied Armies landed in southern Italy at which they fought their way slowly up the Italian Peninsula. In December 1943, the Fifteenth started moving their command, planes and equipment to bases built in the newly secured territory of southern Italy, even though the Allies only controlled about the lower one-third of the Peninsula. This was so we could hit German targets that were out of range of the planes of the Eighth Air Force flying out of England. The Allied ground offensive stalled at the Gustav line in early 1944 because of the winter and the mountainous terrain. We were one the first new units to be added to the Fifteenth.

Our first stop was Waller Field on the island of Trinidad. This flight took us twelve hours, including one stop at St. Lucia Island to refuel. We encountered a mechanical problem, so we had to stay at Waller Field for several days. Our plane ended up parked near a jungle and you could see the monkeys swinging through the trees.

Me at Waller Field, on the island of Trinidad. *Courtesy: John L. Lenburg Collection*

On January 28, 1944, we were off to our next stop, Belém, Brazil. We arrived in a blinding rainstorm. This condition made visibility poor, so Erf had a very difficult time in finding the airfield. Belém sits at the mouth of the Amazon River. By flying at a very low altitude over the mouth of the Amazon, Erf was able to locate the airfield. By the time he landed our plane, it stopped raining. The flight had taken us eight hours and twenty minutes. It was very hot and humid there, since Belém sits only a couple of degrees off the equator. Having spent a very sleepless night, because of the heat and high humidity, we pushed off the next morning for Natal, Brazil.

Belém, Brazil, known for its hot and humid tropical climate, in 1943. *Courtesy: John L. Lenburg Collection*

In Belém, I saw my first airplane accident. An RAF twin-engine plane careened off the runway during takeoff and exploded into a ball of fire. This happened in front of us while we were waiting to take off. The flight to Natal took us six hours. I was excited because my friend Noble was stationed there.

Upon arriving at Natal, I looked up Noble right away. What luck, his outfit would be servicing our plane during the stay here. In the process of servicing our plane, they found a mechanical problem and grounded our plane until repaired. The part needed was flown in from the States to make the necessary repairs. I was able to spend five days there. We were restricted to the base during our stay. I was able to get into the town of Natal several times with the help of Noble and some of his buddies. Since I was posing as a person from Noble's outfit, I did not try my luck for a third time.

The next leg of our journey would take us over the Atlantic Ocean. We were to fly from Natal to Dakar, French West Africa (now Senegal).

Our instructions were to take off at three-minute intervals and fly alone as we had been doing. Since we would be flying at a lower altitude, we kept a look out for German submarines. One had surfaced and fired on a plane the day before. They knew that American warplanes would use this route to go overseas. On February 2, 1944, we arrived in Dakar. The flight took ten hours and forty minutes and we encountered no problems.

At this facility, we were required to provide an overnight guard for our plane. I drew the short straw on this assignment. It was an interesting experience, though. The air base assigned a Senegalese soldier to stand guard outside each aircraft but one of our crew members had to stay inside. Once the sunset, we were under orders not to come out of our aircraft. When I awoke the next morning, there was this Senegalese soldier standing outside of our plane holding a long rifle with an attached bayonet. He wore short khaki pants with a red fez on his head and no shoes. Smiling, he saluted me and said, "Good morning, Sahib."

On February 5, 1944, we took off and flew over the vast Sahara Desert to Marrakech. The view of the desert from the air was bleak. We did pass over several camel caravans crossing the desert. The flight took us seven hours and thirty minutes. There were at least some palm trees around Marrakech. Upon arriving at the air base, we received the order to carry our forty-five-caliber pistols with us. Some kind of an uprising was underway.

Some of us were able to make a trip into the Casbah during our stay there. Now this area was off limits. Since we had always heard about it, curiosity got the best of us. We found a driver of a horse-drawn carriage, who told us that he would take us in. Once inside, we encountered a problem. The carriage hit a little Arab boy, so the driver had to stop. Of course, we were starting to draw a crowd of Arabs right away. Everyone began shouting and screaming. It started getting a little hairy and I thought that maybe we would have to shoot our way out of it. Finally, the driver cracked his whip on the horse's rear end, the horse bolted forward, and the people moved out of the way. I might add the little boy never sustained serious injuries; instead, he was just shaken up.

We hung around Morocco for a few more days, and then got ready for the next leg of our journey, Tunisia.

On February 7, 1944, we left for Tunis, Tunisia. We were to land at Ounda 1 Air Base, a captured German air base nearby. It was a welcome sight to see the green vegetation of northern Algeria as we flew along the Mediterranean Sea. The closer we got to Tunis, the scenery started taking on a different look. It started looking more like a war zone with burned out tanks, armored personnel carriers, trucks lining the roads and a few crashed German aircraft on the beach. Our flight to Tunis took us over the famous Kasserine Pass. It was here that U.S. troops fought Rommel's Panzer Tank Corps from February 14 through February 22, 1943, to try to evict them from North Africa. Hundreds of burned-out vehicles and tanks littered the desert floor from the battle that had taken place there. Nearly 30,000 Americans took part in the battle and one in four ended up as a casualty. The flight from Morocco to Tunis had taken us seven hours and fifteen minutes.

Wheeler and Troy in Tunisia. *Courtesy: John L. Lenburg Collection*

Our stay at Ounda was expected to last a week due to the fact work on the air base in Italy had not been completed. The weather was on the chilly side and the wind blew constantly. There was an old Roman aqueduct within view of our field. Since we were housed in tents, we decided to warm our tent up a little bit, by making a heater. Our "heater" was made of a fifty-five-gallon drum cut in half and a chimney made out of large metal cans that we obtained from the mess hall. Our fuel reservoir was a deicer tank that we had picked up at an airplane dump nearby. After filling the tank with German aviation gas, we were ready to light the fire. Unfortunately, we almost burned our tent down. We forgot to attach some kind of a regulator to regulate the flow of gasoline. After adding a regulator (which consisted of a piece of rubber tubing inserted in the gas line with a clamp on it) and replacing the tent pole, we were able to move back into the tent. Now it worked great and we had heat. We had to constantly keep watch on our tent area since the Arabs would slip in and steal our mattress covers or anything else that they could get their hands on. The covers they used for clothing. In Tunis, we practiced some formation flying during our stay. I was able to get into the city of Tunis several times.

February 19, 1944: Arriving at Spinazzola Army Air Base in southern Italy, that became the base camp for the crews of the 460th Bomb Group. *Courtesy: John L. Lenburg Collection*

- 1944 -
ARRIVING AT SPINAZZOLA

On the morning of Friday, February 18, 1944, the day of my twentieth birthday, we received word that we were leaving for our newly completed air base near Spinazzola, in southern Italy.

By the time the ground echelon of the 460th had arrived at the base, they had assigned our group to the 55th Bomb Wing, made up of six bomb groups. We spent most of the day getting things ready for the last leg of our long journey.

On Saturday morning, we loaded our gear into jeeps and rode out to our planes with Lt. Roy Hansen's crew. The flight engineer of the crew was S/Sgt. R.B. Frazier. Frazier was a good friend of mine. In fact, our crew was good friends with the rest of Hansen's crew since we had gone through training together. We rode together in the jeep to our planes. All the while, we were laughing and joking with each other as we rode.

Since our planes were close together, we unloaded our baggage and got our plane ready for our trip to Italy.

Since we would be entering a war zone, fifty rounds of ammunition were pre-loaded in our guns. We had a high overcast day, but the weather was not too bad for flying. As we approached the coast of southern Italy, the weather got worse and the overcast closed in. Consequently, the weather conditions got worse and we lost sight of some of the other planes flying in our formation. Southern Italy is very mountainous, so Erf dropped the plane down closer to the ground. With the cloud cover so thick, he dropped to 500 feet below as we flew into the valley below. Erf picked up the radio transmission from our control tower, "Dolly Tower," at our destination's airfield. Suddenly the valley came to a dead end and we had to pull up and reverse our course. We knew we were in a bad situation. Eventually we found a railroad line and followed it until we came across a couple of small railroad stations. The signs on the stations read, "Gravina" and "Altamura." By doing this, Erf was able to find our air base. Our base sat in a valley surrounded by mountains in the distance and the railroad ran right by it.

It was not until the next morning that we heard that Lieutenant Hansen and Lt. Norman E. Corey's planes crashed into these mountains. This was our group's first casualties, and some were my best friends. I guess in the excitement of unloading our baggage at the flight line before taking off, one of my bags got loaded on Lieutenant Hansen's plane. Of course, I lost everything. This hit our crew pretty hard. We had just arrived at our destination and already we needed two replacement crews. Now dying became a reality. Because of the heavy snow in the mountains, they could not recover the bodies of the crews until late spring.

To arrive at our final destination, we had flown more than 11,000 miles and it had taken us more than sixty-four hours of flying time. The base consisted of a 6,000-foot runway, with sixty-four steel hardstands and a taxi strip located on a 1,000-acre wheat farm. Also, a

Ground view of Spinazzola Army Air Base, Italy, 460th Bomb Group, with the B-24 "Hairless Joe" in the background. *Courtesy: U.S. National Archives and Records Administration*

vacant farmhouse used as group headquarters and a large granary barn for our briefing quarters. Another granary barn was transformed into an infirmary and a large hay barn into a motion picture theater.

Our crew, except for the officers, lived out of a tent in the 760th enlisted men's area and the officers in their own dedicated tent in the officer's area. Our tent was the last in the enlisted men's area of 760th Squadron. We were next to the road that ran into Spinazzola and a railroad ran parallel to the road. It was cold and started to rain again. This condition made our base a sea of mud. (It was hardly the "beautiful, warm sunny Southern Italy," I had read about in travel brochures.) The sides of our tent were flapping in the cold wind and there was no heat. There were no kitchens to prepare our food or mess halls in which to eat. So the food had to be prepared outside and everyone had a choice of eating their food by sitting on the ground, standing in the rain or going back to their tent.

Since this was going to be our home for who knew how long,

Home, sweet, home at the 460th Bomb Group camp. *Courtesy: John L. Lenburg Collection*

Who says it doesn't snow in southern Italy? *Courtesy: John L. Lenburg Collection*

How camp looked months later in the spring of 1944. *Courtesy: Duane "Sparky" Bohnstedt*

460th Bomb Group Headquarters, 1944-1945. *Courtesy: Duane "Sparky" Bohnstedt*

Me feeding the neighboring crew's monkey in camp in southern Italy. *Courtesy: John L. Lenburg Collection*

we decided we had better start making it livable. So the first order of business was to get some heat. They had another stove made for us, but we did not make the same mistake as we did in North Africa. This time we had to install a regulator on the gas line. The following weeks we spent making our base more livable, getting the planes ready for combat, and practicing close formation flying. We eventually put a tufa blocks floor in our tent. Some crews even put up sidewalls or built small huts out of the blocks for their living quarters. As the days and weeks went by, conditions gradually improved.

On March 8, 1944, we went on a training flight. Suddenly developing engine trouble forced us to land at an air base in Cerignola. We stayed there several days until they repaired our plane. The bomb group stationed there went on a bombing mission the next day. It was here that we got our first taste of what to expect when going out on a combat mission. We were able to see the planes returning after completing their mission. Some shot up had wounded on board. A few others did not return at all.

On a free day, visiting with some Italian kids in Bari, Italy (back row, left to right): Troy, me, Pappy, and Jack. *Courtesy: John L. Lenburg Collection*

-1944-
FLYING MISSIONS

On March 19, 1944 our group became active. Everyone started sweating out what the next day would bring. The 460th's first mission was to bomb the marshalling yards at Metovic, Yugoslavia; Colonel Crowder led the mission. They never listed our crew on the mission schedule posted on the squadron bulletin board. We were all up early the next morning and watched the group take off. It was supposed to be a milk run; in other words, not much enemy resistance was expected. The group returned in the afternoon and we had suffered no losses. But the colonel was furious. The formation flying was terrible, too spread out. He pulled our group out of combat readiness until the group's formation flying improved. Therefore, the next ten days, the group spent practicing formation flying.

In the Mediterranean theater of operations, aircrews had to put in

A B-24 Liberator of the 460th Bomb Group in action. *Courtesy: U.S. National Archives and Records Administration*

fifty combat missions before being eligible to be rotated back to the States.[2] Something many of us hoped happened, for it meant we would be able to return home. Any mission that crossed the 45th parallel counted as two missions completed. Those usually flew to Austria, Rumania, Hungary or Germany. Since our planes were not pressurized, we carried oxygen tanks and wore oxygen masks. The mode of dress for these missions was whatever it took to keep warm. It was extremely cold at the altitudes that we would be flying at with temperatures of 30 to 40 degrees below zero.

My clothes and equipment could not be too bulky, or I would not be able to move around the plane or get into my turret. Most of our crew did not wear the seat parachute, except the pilot and co-pilot. The seat chute was too big and bulky. We used a chest chute, consisting of a parachute harness that snapped into two large rings on the front of the harness.

---

[2] *Later, a tour of duty changed from 50 to 30 missions. Each combat sortie was counted as single missions.*

Under my parachute harness, I wore a "Mae West," an inflatable life jacket (so named because of the generous bulge it gave airmen around their chest that was reminiscent of physical endowment of the famous actress), in case we made an emergency landing in water. Usually, I left my chute pack near the rear bulkhead on the flight deck and just wore the harness. I wore long johns or a quilted flight suit under my coveralls, jacket and G.I. shoes. Over all of this, I had to put on a flak suit and then be able to squeeze into the turret. I wore a sheepskin-flying helmet with earphones and a flak helmet over that. Some of the crew carried a forty-five-caliber pistol in a shoulder holster but I elected not to carry mine with me. If shot down and captured that could give the enemy a good excuse to shoot you on the spot. They could say you were a spy after parachuting from the sky above.

The flak suit, made of overlapping steel shingles in a quilt-like cover, was designed to protect the vital areas of the body, since 79 percent of all wounds occurring among heavy bomber crews were from low velocity shell fragments, from exploding anti-aircraft shells. A new flak helmet that Maj. Gen. Malcolm E. Grow developed tested with excellent results. It was a flexible five-piece close-fitting helmet weighing three pounds.

I have included a summary of the missions on which I flew. I will try to recap of each mission from the notes and copies of reports that I have. For those killed that I knew, I would write his name down but not the mission. The aircrews were broken into several groups, so you did not fly every mission that was scheduled. When I flew my twenty-seventh mission but credited with my thirty-sixth, the group was flying its fifty-third mission.

Many times, a badly damaged aircraft would make it back to the base, only to crash in attempting to land. Some planes were so badly damaged they never flew again. At times, a plane would have an accident and pile up on takeoff. Therefore, all our losses were not always due to enemy action.

After completing a mission, every crew had to go through a debriefing session. Each crew would tell what they observed on the mission. The Red Cross gave us coffee, doughnuts and a shot of whiskey. I was not a coffee drinker and after being on oxygen for five or six hours, I never drank the whiskey. If I did, I would most likely pass out.

March 30, 1944

On the evening of March 29, 1944, an officer posted on the bulletin board outside of the orderly room a list of the crews flying the next day's mission. Our names were on it. They scheduled our crew for the group's sixth mission. We would not sleep well this night.

At 5 A.M., in the darkness of early morning, came the wakeup call from Sgt. Bob Cutler. Of course, everybody groaned and moaned as we rolled out of our cots. After dressing and brushing our teeth and shaving (that is if you needed to) and the rattling of mess kits came breakfast. We ate breakfast on the ground or back in our tent,

The barn in which our briefings were held. *Courtesy: John L. Lenburg Collection*

Briefings entailed the squadron leader giving last-minute tips and advice to the crew. *Courtesy: U.S. National Records and Archives and Records Administration*

because the mess hall had yet to be constructed. Early morning church services followed for those wishing to go.

Then it was off to the group's briefing room held in a barn. Once inside, all you heard was a loud murmur, since everyone seemed to talk at once. Of course, we were all excited and nervous as we waited anxiously for the arrival of the briefing officer. Upon his arrival, a sudden hush fell over the room as he pulled back the curtain covering a large map posted on the wall in the front of the hall, showing the route taken for that day's mission. If it was a milk run, there would be a sigh of relief but more groans if it was a tough one. Today, our first mission was a milk run. Our target was a German airfield at Mostar, Yugoslavia. We did not expect to encounter much enemy resistance on this mission.

At the finish of the briefing, we were loaded onto jeeps and trucks and taken out to the flight line. At the flight line, we were given parachutes, Mae Wests, flak vests, flak helmets and other equipment and then it was off to the hard stand where our plane was parked. The planes looked cold and lifeless standing there in the in the early dawn. Next, we did a visual pre-flight check of the plane's exterior. First, we checked gas and oil tanks to make sure they were properly sealed. Next checked the landing gear, superchargers, wings, and tail assembly.

Dressing for sub-zero temperatures. *Courtesy: Courtesy: U.S. National Archives and Records Administration*

Once finished, we boarded the plane to start a pre-flight cockpit check of gauges, electrical equipment and controls. After this was completed and all of our equipment was onboard, we watched for the firing of starting flare from the Dolly Tower (the control tower). If the flare was red it meant a "stand down," but if it was green it was a "go."

B-24 standard issue worn in combat—a modified steel helmet and flak jacket. *Courtesy: U.S. National Archives and Records Administration*

Our group was the only one carrying out this mission. That meant that there would be about thirty-to-forty aircraft participating in the mission, not hundreds as you might think. Colonel Crowder would lead the mission.

At 0927 hours, the green flare fired from Dolly Tower, followed by the "put-puts" sound of the engines starting up in each plane by the engineer. This auxiliary power unit provided electrical power for the plane until its engines were running. Following the groans and coughing puffs of smoke from the engines, each one came to life. The engine's exhaust began to permeate the air. Our crew chief Philip Baideme stood by with a fire extinguisher in hand in case of a fire. There was a full throbbing of the plane when all four engines were running. The once tranquil airfield suddenly sprung to life as the planes with a full load of gasoline and bombs began rocking on their shocks with engines running full blast. The planes pulled out on the taxi strip lining up for takeoff. It was quite a sight to see forty planes lined up with their engines running, waiting to take off.

"Dolly Tower," the control tower at our destination airfield on B-24 missions. *Courtesy: John L. Lenburg Collection*

After the signal was given from Dolly Tower, there was a sudden roar of engines and the lead plane rolled down the steel-matted runway. When our turn came, we pulled out onto the runway. Thirty seconds after the preceding plane took to the sky, we received the go signal from the control tower.

Erf always chewed a stick of gum when he flew. I am not sure he did this to ease the tension or to clear his ears. Maybe it was for both. When things got exciting, Erf would really work that piece of gum over. Chewing his gum at forty miles a minute, Erf put the flaps down and set the props at full pitch. As he pushed the four throttle levers all the way forward, our plane lurched forward, picking up speed very fast as we headed down the runway. Since Erf and Alan were busy at this time, I stood between them. I assisted them in keeping check on the many instruments to make sure all were functioning okay. You could hear the strain of the engines thrusting forward as we went faster. We were past the point of abort.

Soon the wheels began bouncing a little as the shocks began to stretch out. Our plane began to lift from the increased speed. As the end of the runway drew closer, Erf eased back with the control column. We either lifted off or became a big puff of smoke that I had seen happen on several occasions. Fully loaded, our plane, which weighed close to thirty-five tons, strained and groaned as it lifted off the runway, barely clearing the field at the end. Wheels and flaps went up and we gained altitude to begin the tedious task of rendezvousing with the forty-three planes in our group. We kept circling over Altamura until all planes were in formation. There was always the threat of colliding with another plane. I put on my flak suit and helmet and climbed into my position in the top turret. At 10,000 feet, the oxygen mask went on. The weather was bitter cold, so I started occasionally test firing my guns. This was to make sure they have not frozen up.

Once in my turret, I found my head was low, so I did not have much of a view of things below and parts of my turret blocked my view when looking sideways. Apart from this, the visibility was excellent since my Plexiglas top was frameless. There was a footrest under the turret upon which you could rest your feet. The sun shining on the curves of the

Plexiglas created a harsh glare, which was a nuisance once in flight. With the guns so close to your head, the noise and kick when fired was terrific. Smoke tended to accumulate in the top of the turret and thus causing a problem. It was important to keep the bomb hatch door closed, so that spent shells from my guns would not fall into the bomb bay. This would cause the bomb-bay doors to jam.

The formation flying for the mission was excellent and it took us four-and-a-half hours to complete. We hit the target with 500-pound bombs from 20,500 feet and had light to moderate flak. I filled out our flight log and from that point on, never looked back. Our group was fully operational again.

April 3, 1944

My next mission was a marshalling yard at Knin in Yugoslavia. This was a four-and-a-half-hour mission and again we encountered light resistance from the enemy. Thirty-four planes hit the target with 500-pound bombs from 20,500 feet. On many of the raids, the Army Air Force would send a much larger force to a target. This would draw the German Luftwaffe to defend it, while a smaller force hit a different target in the same general area. A pattern started to emerge of how we would be flying. It looked like we would be flying about every third mission.

April 6 and 12, 1944

The next mission was an airfield at Zagreb, Yugoslavia. We again hit our target with cluster fragmentary bombs from 20,500 feet. This mission lasted five hours. In fact, on April 12, 1944, we went back there again with a smaller force to bomb the airfield and marshalling yards, while a larger force raided Wiener Neustadt in Austria. This larger force drew most of the German fighters, Messerschmitt Bf-109s, often called Me-109s, and Focke-Wulf Fw-190 Würgers, both single-engine, single-seat aircraft. The raids our group was participating in were getting bigger.

Hitting bigger targets meant more planes flying together. This would translate into heavier enemy resistance. I guess that we were easing into taking part in larger missions. It was on this mission that we started dropping chaff as we neared a target. Chaff was small pieces of tinfoil that our waist gunners would dump out to foul up the German radar.

Hits were observed on the northeast portion of the field, on the taxi-strips and across into the marshalling yards. Two large explosions in the hangers were visible.

April 15, 1944

My fifth mission was a raid on Bucharest, Rumania. Our commanding officer, Colonel Crowder, was to lead our group. He was the lead plane in the high box. We were flying second to the lead in the low box, carrying 500-pound bombs, on this mission. At briefing, they told us to expect enemy fighters and anti-aircraft gunfire that day.

As we neared the target, enemy fighters attacked us. They came from twelve o'clock and out of the sun. The first plane was an Fw-190, followed by eight Me-109s that passed through our formation. Of course, all of the guns, brought to bear on them, were blazing away at the Fw-190. Following him closely were two Me-109s who seemed to be firing at the lead plane, piloted by Colonel Crowder. Flames erupted in the bomb bay of the plane and then there was an explosion. The plane pulled up in a stall, half rolled, and then spiraled down to earth. Two parachutes opened and that was all. One of the rituals, when one of our planes went down was to count the parachutes.

The fighters that hit us were from Hermann Goering's elite yellow nose fighter group. These planes had the nose of their propeller painted bright yellow. We encountered light to moderate flak over the target and returned to our base back to Italy. Killed in action was Colonel Crowder. Thereafter, Col. Ben Harrison assumed command of our group. The Fw-190, which went through our formation, was a decoy to draw our fire. It was the first plane that I saw go down due to enemy action.

We had hit the target that day from an altitude of 20,000 feet. Thirty-two aircraft from our group dropped forty-eight tons of 500-pound bombs directly on the target to obliterate it. The 15th Air Force lost fourteen planes and shot down thirteen enemy fighters that day. Flying time for this one was seven hours.

Now reality started to set in with me: I could be killed. This was only the beginning, since there would be many more. With losses averaging ten

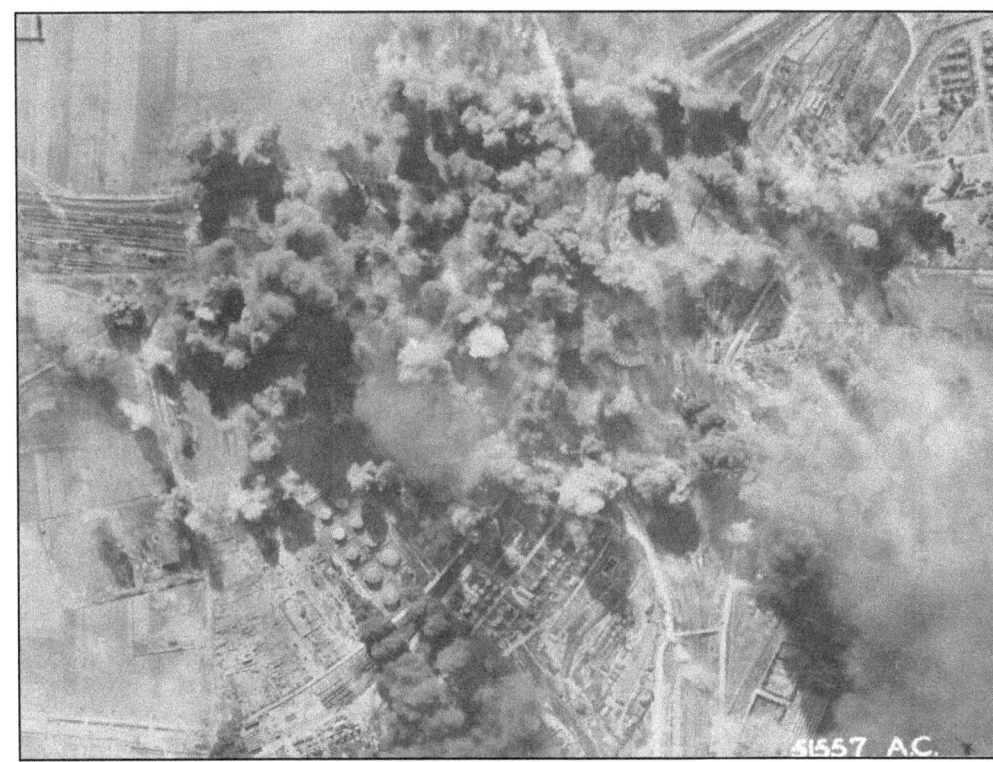

Targets are destroyed by a flurry of 500-pound bombs dropped on them. *Courtesy: U.S. National Archives and Records Administration*

percent, a crew could expect their luck to run out after ten sorties. Anything over ten missions, you would be living on borrowed time. Our attrition rate was fairly high back then, so by the time I had reached my twenty-fifth mission, we had lost quite a few of the original crews from our squadron many with whom we had trained. The following are the names of some of those killed that I knew:

Col. Robert T. Crowder	Sgt. Walter W. Rowe	S/Sgt. A.J. Horn
1st Lt. Firman E. Susank	F/O B. Trice	S/Sgt. Cecil R. Boles
1st Lt. Stevens	S/Sgt. Raymond B. Frazier	T/Sgt. Justin D. Danner
1st Lt. Hansen	S/Sgt. John H. Blake	Sgt. Horace C. Barksdale
1st Lt. Jack W. Smith	S/Sgt. Vincent L. Healy	Sgt. Donald M. Eggleston
2nd Lt. Miller	S/Sgt. Harry P. Hoyer	Sgt. Clayton E. Sprague
2nd Lt. Gluntz	T/Sgt. Guy Marsh, Jr.	Sgt. Loren Zimmerman
2nd Lt. Norman E. Corey	S/Sgt. Louis O. Moss	Sgt. Robert K. Williams
2nd Lt. Robert G. Evans	S/Sgt. Edward H. Newsom	Sgt. Elmer H. Thomas
2nd Lt. Gifford	S/Sgt. Walter E. Erickson	Sgt. William C. Trego
2nd Lt. R.H. Woods	S/Sgt. Charles B. Dorfman	Sgt. C.A. Shaeffer

The replacement crews that arrived brought silver colored B-24s; the original planes, such as ours, were ones painted olive drab. It was not long before there were not too many olive drab painted planes flying in our group. As we lost more and more of our friends, whom we had gone through training with, our crew became more isolated. We were reluctant to become too friendly with any of these replacement crews.

By the middle of April 1944, life at our base started improving. The 760th Squadron built and opened enlisted men's mess hall. Opening of mess halls for the 761st, 762nd and the 763rd and one for the officers followed this. Italians that were hired did KP duty. There was a great improvement in the food served to us. Many of the crews had finished replacing their tents with small shelters built out of tufa blocks. With all of this construction, the base began to look like a small city. In some areas, street signs began sprouting up, like "Milk Run Alley" and "Stagger Inn Road." The last one was because they were always having parties at their huts during their free time. We even began to watch movies shown in our converted barn theater. I found an Italian family who would do our crew's laundry. We also traded our cigarettes and chocolate bars to the Italians for fresh oranges and eggs. The eggs we would cook on our homemade the stove in our tent.

Anti-aircraft guns, manned by British anti-aircraft gun crews, protected the airbase. Our crew made friends with a crew that manned the anti-aircraft gun batteries near where we parked our plane. They would invite us over for a spot of tea occasionally. After several months, we became good friends with this crew. We invited them over to our tent for several parties with our crew. They always brought a keg of wine, Canadian Black Horse Ale and sometimes a fifth of scotch. It would take several days to air our tent out after one of these parties. Our crew also made friends with some members of a South African tank corps, pulled back from the front lines for a rest. We had several parties with these guys also. For some reason all of these get-togethers were held in our tent. Maybe it was because our tent was the last one and was by the road. I went on several daytime excursions into the Italian Mountains with the British when I had time off. We would ride in a British lorry.

April 20, 1944

The next mission, our primary target was the marshalling yards at Castle Maggiore, five miles north of Bologna. This was to be a milk run. I was to fly with Lieutenant Schuneman's crew as a replacement, since his engineer was ill. On our way, the plane developed engine trouble, so Schuneman and I decided that it would be best to abort. If you aborted a mission, you had better have a good and valid reason or you could be court marshaled. The primary target was not hit because of overcast skies. They hit a target of opportunity, the rail bridge at Fano, Italy.

April 21, 1944

My eighth mission, we were off to hit the marshalling yards at Bucharest, Rumania, part of the Eastern front. This was the group's thirtieth mission. Flying our plane was our squadron's commanding officer, Maj. Bob Martin. Erf served as co-pilot. The target area had become clouded over, so we hit the secondary target. This was the marshalling yard at Turnu-Severin. Again, we used 500-pound bombs and flew at 21,500 feet and the mission took almost seven hours. Flak was fairly heavy. We lost one plane. One was damaged by flak and another ditched in the Adriatic, near the Island of Vis, after running low on fuel.

April 24, 1944

Three days later, we bombed the marshalling yards at Bucharest again. We carried 500-pound bombs. Thirty-three of our aircraft hit the target with thirty-eight tons of bombs from an altitude of 23,000 feet. The marshalling yard observed was very heavy with rail traffic. A heavy concentration of hits covered the entire width of the marshalling yard. Hits were also visible near the roundhouse. Going to the target, we flew over an airfield where we also spotted sixteen enemy fighters and fourteen enemy transports and gliders.

Again, we knew the mission would be a little rough after our last experience in this area. We had fighter escort by P-47s, P-51s, and P-38s but the enemy still hit us with Me-109s and Fw-190s. One attack by a Me-109 came from one o'clock high and then broke left at about 150 yards. Two others attacked from four o'clock low and the other came in level. They closed in to about 200 yards when one peeled right and the other left. Four other attacks came from enemy fighters from 6 o'clock level. Another two-enemy aircraft hit us from three o'clock level, but they peeled off out of our range. Ten other enemy fighters flew high and left of our formation at eleven o'clock but then broke away to the right. The attacks were not very aggressive.

When we flew missions to major targets such as Vienna, Bucharest, Munich or southern France, we had fighter escorts. In most cases, they were effective in keeping the enemy fighters at bay, but you have to keep in mind our bomber formations consisted of 500 to 800 planes. These

B-24 bombers come under heavy German artillery 88-millimeter anti-aircraft fire.
Courtesy: U.S. National Archives and Records Administration

formations of bombers spread out many miles, so it was a very difficult job for our escort to protect all of us. Of course, there were the times when one group of enemy fighters would engage our fighter cover, while another group of enemy fighters would hit us. They would be like a swarm of bees when they hit you. Usually, they concentrated their fire on one plane. If they crippled the plane to the point that it could not keep up in formation, they would finish the job as the plane straggled behind. Flak and enemy planes damaged seven of our aircraft.

As we approached the target area, fire from the anti-aircraft guns would get very heavy. The Germans would throw up what they called a box barrage. Once you were on your bomb run, there was no deviation from your course. At this point, the bombardier would control the plane with his bombsight. Since we were flying at such a high altitude, the slightest deviation could cause the bombs to miss the target by several miles. Therefore, we had to fly into the flak the German guns would put up and hopefully through it.

At times, the Germans would lay a carpet of flak so thick that it looked like you could get out and walk on it. When this happened, a part of Alfred Tennyson's famous poem, "The Charge of the Light Brigade." would come to my mind:

> Theirs is not to reason why,
> Theirs is but to do or die:
> Into the valley of Death
> Rode the six hundred.
>
> Cannon to the right of them,
> Cannon to the left of them,
> Cannon in front of them
> Volley'd and thunder'd;
> Storm'd at with shot and shell,
> Boldly they rode well,
> Into the jaws of Death,
> Into the mouth of Hell
> Rode the six hundred.

We also had sort of a saying amongst ourselves as we approached a target and saw the puffs of black smoke from the barrage of flak ahead: "It's not the black puffs of smoke that you see that hurt you but the ones that you don't see."

Enemy fighters would also wait as we came out of the target area looking for planes that had been hit and could not keep up with the formation. As the anti-aircraft shells exploded around us, it would sound like someone was throwing a handful of gravel at the plane. Then there were the times when the enemy got a direct hit, which might cause the plane to explode in a ball of flames. As we would leave the target, the pilot would ask for a damage assessment.

April 29, 1944

By the end of April, I had chalked up almost a dozen missions. The next one was the submarine pens at Toulon, France. Nearly 500 heavy bombers of the 15th, with fighter escorts of P-38s, and P-51s, hit the target area. My view of this vast armada of planes from the top turret was an awesome sight. It stretched out as far as the eye could see.

We encountered heavy flak and a very aggressive force of Me-109s, Fw-190s and Messerschmitt Me-210s built by the Germans to replace their Bf-110 heavy fighter. The Me-109s attacked coming in at four abreast. The 15th shot down seventeen enemy fighters and we lost twenty-eight of ours on this mission.

For this mission we used 1,000-pound bombs and hit the target from 20,500 feet. Nine of our aircraft emerged damaged. The mission lasted more than eight hours and the missions seemed to be getting longer. At the outset of the mission, we saw two submarines diving outside of the harbor. We also spotted two battleships, three cruisers and six destroyers mooring in the harbor, and six cruisers in the Gulf of St. Tropez.

For some reason, our engines were burning fuel up faster than usual. I told Erf that I did not think that we would make it back to our base, so he decided that we should land at a B-25 medium bomber airfield on the island of Corsica. We encountered some problems on landing since the runways are much shorter in length on a medium bomber airbase. After refueling, Erf taxied the plane to the runway for takeoff. People lined the runway to watch us take off. Erf said they thought that we would crash and were going to watch us try to take off on this shorter runway.

Erf set the brakes, revved the engines up to full RPM, and gave it full flaps. The whole plane was shaking and vibrating so bad that I thought every rivet was going to pop off the plane. Releasing the brakes, we shot down the runway. Erf only used about half the runway before we were airborne.

May 2, 1944

The 460th and the 464th were to hit the marshalling yards in Parma, Italy while other units of the 15th Air Force were bombing targets in northern Italy. What started out as an easy mission turned almost catastrophic for us. Our group was to fly up the West Coast of Italy, turn inland and bomb the marshalling yards in Parma, and then proceed across Italy to the Adriatic Sea. After turning south, we were to continue past the front lines, turn inland and land at our base.

Everything was going along fine, until we turned inland along the East Coast. Our formation was flying at 21,000 feet and above the overcast, approaching our Initial Point or I.P. on our flight plan where our bombing run was to begin, when it became obvious that the overcast sky over target area would obscure our view. Colonel Harrison led our group. He decided

May 2, 1944: Flying up the Adriatic Sea to Yugoslavia. *Courtesy: John L. Lenburg Collection*

Me and Jack after the May 2nd mission.
Courtesy: John L. Lenburg Collection

Where the 40-millimeter hit our plane. *Courtesy: John L. Lenburg Collection*

to letdown or descend the aircraft and attack the target under the overcast. The letdown started at the I.P. The bomb run started as we were dropping down through the overcast. Two planes, to our immediate right, collided. The wing on one of the aircraft buckled and then both planes went down in a ball of fire. Five chutes opened, coming from one of the aircraft. This looked like one of those missions where nothing was going to go right. We lost all visual contact with the other planes in our box. Consequently, we were not able to drop our load of bombs on the target.

Erf continued flying by instruments until we suddenly got a break in the visibility. We found that we were over water, with land coming up ahead. Our navigator, Hendricks, told Erf that we were approaching the West Coast of Yugoslavia. Erf felt certain we were approaching the East Coast of Italy after turning the aircraft and heading south. We finally agreed that it was the coast of Italy, but it was unclear if we were above or below the front lines. Erf jettisoned the bombs into the Adriatic Sea. Incidentally, we were flying alone and unable to locate the rest of our group. Radio contact was very limited since the enemy could pick up our radio signals. Erf had followed the flight plan as told at the morning briefing.

We spotted a small coastal village as we proceeded south. Hendricks said that we had crossed the battle lines, but Erf did not think we were that far south yet. Erf decided to see if he could determine just where we were. He swung the plane around and headed for the village, dropping down to about 500 feet. He asked me to drop down out of my turret to help them.

We approached the little coastal town looking for some clue as to what our actual location might be. Mike Brown, who had come up from his position in the nose, and I were standing between Erf and our co-pilot Barrowcliff. As Erf banked our plane around to go back out to the Adriatic, Mike and I looked to see what landmarks we could find. All of the houses appeared boarded up and nothing was stirring in the streets. Erf decided to make another pass, as another stray B-24 we had picked up, followed us.

As Erf banked the plane, all hell broke loose. There was an explosion on the flight deck and anti-aircraft shells were bursting all around us.

Mike, who was standing beside me, spun around and fell to the deck bleeding profusely from the head. I started first aid on Mike, who appeared to have suffered a scalp wound on his forehead. Erf pushed the throttles to full open to get us out of there. He worked the plane's flight controls, so our plane would slip to the right. This made it harder for the Germans to get a beat on us. By doing this, the anti-aircraft shells kept exploding to the left of us. The tracers from their machine guns kept going by us until we got out of their range. They had hit our number three engine and Erf had to feather the prop (shut it down).

After getting Mike's bleeding under control, I assessed the damage. No one else was injured. However, the machine guns and 40-millimeter anti-aircraft shells the Germans had been firing on us left our plane riddled with holes. We also had a large hole next to the radio operator's station from the exploding shell that had hit on the flight deck. Part of it went out through the top of the aircraft next to my turret. If I had been in my position, most likely I would have lost both of my legs. I guess now we knew for sure which side of the front lines that we were over. The other plane that was following us pulled up short and headed back out to sea.

We limped down the coast until we made radio contact with our base. As we neared our base, I fired a red flare. The tower cleared us to make an emergency landing. As our plane rolled to a stop, an ambulance and flight surgeon met us. Erf's stick of gum got a workout that day.

Rumor had it that all personnel of the 460th were going to be restricted to our base. They turned out to be false. We were going to have to fly two missions a day, in support of the upcoming invasion, and then the 460th relocated.

May 6 and 12, 1944

On May 6, 1944, we were to hit the marshalling yards in Craiova, Rumania, a highly industrialized region known for its agriculture. Our plane developed a mechanical problem and we had to abort our flight. At the May 12 briefing, it was announced that the Allied Armies in Italy were about to launch their biggest offensive of the war. The Allied Armies advance stalled below Rome, along the Gustav line at Mt. Cassino. Our assignment was to bomb the marshalling yards at Ferrera, Italy as our target. Twenty-one groups of heavy bombers and six groups of fighters took part in this mission. This amounted to well over a thousand planes. The mission took seven-and-a-half hours and our group of thirty-two aircraft dropped sixty-four tons of 500-pounds bombs. Flak was light, four aircraft damaged, but no enemy fighters encountered. Two merchant vessels sighted off the beachhead burned out of control.

May 14, 1944

At 0840 hours, thirty-six aircraft from the 460th took off to bomb the marshalling yards at Treviso, Italy, with our squadron C/O Major Martin, flying our plane. The group formed into two attack units, with us flying the lead. We rendezvoused with the other groups in the 55th Bomb Wing between Spinazzola and Marghareta. We led the wing, which consisted of five bomb groups on this mission. From 21,000 feet, we dropped 500-pound bombs on the target. The mission took five-and-a-half hours, producing excellent results.

May 17, 1944

Our next target was a marshalling yard in Piombino. We had fighter escort to the target area. Our group dropped eighty-four tons of 500-pound bombs on it, from an altitude of 22,000 feet. We encountered some damage from flak.

May 19, 1944

The Allied ground offensive in Italy had started. We were bombing anything and everything to disrupt the German transportation system, including train yards and other targets. On May 19, we hit the marshalling yards at Bologna with twenty-nine aircraft in two attack units. We had P-38 fighter escort and dropped seventy-three tons of 500-pound bombs on the target. We encountered some flak over the target area.

May 22, 1944

My next mission was Valmontone, Italy. The Americans started a push to try to break out of the Anzio beachhead. This was to be a major show of force by the 15th. We were not able to drop our bombs on the target, due to bad weather conditions in the region. Eleven aircraft suffered flak damage.

We also observed no enemy rail traffic between Bologna and Rimini. Everything was bottled up.

May 27, 1944

On May 27, we hit the marshalling yards in Nimes, France, with sixty-three tons of 500-pound bombs. This was in preparation for "Operation Overlord" or "D Day." We encountered some enemy fighters and lost one of our planes, even though we had fighter escort, and down one enemy fighter in the process. This mission took eight-and-a-half hours to complete. We spotted a heavy cruiser and submarine off Cape Juan.

May 29, 1944

At 0630 hours, thirty-six aircraft of the 460th took off to join the 15th Air Force to bomb targets in the Vienna area. We were to hit the German aircraft factory at Atzersdorf, Austria, which was in this area. It was part of a total effort of the 15th to cripple the German aircraft industry and airfields. This Vienna area was the second most heavily defended area in Europe. Missions to this area had been very costly in men and planes. The Germans had a very elaborate aerial defense system.

The area was divided into three flak districts: Wiener Neustadt, Vienna, and Moosbierbaum. It was known as "flak alley" because of hundreds of anti-aircraft guns that the Germans used to defend the area. The flak would be so thick that it looked like you could get out and walk on it. The Jerries also could put up to 1,000 fighter planes for their defense. We had P-38s and P-51s fly fighter cover for us. The sight of hundreds of aircrafts leaving their vapor trails was overwhelming.

About half way to the target, I test fired my guns and they started to run away. This sometimes will occur and is due to the mechanism freezing up due to the sub-zero temperatures. When this occurs, your guns will stop firing when you release the firing switch but the ammo feeding mechanism will not stop. Consequently, the ammo would jam up the top of the turret.

I really had to work very hard in the small space in which they confined me in order to correct the problem. Since we had not reached our target yet, I needed to remedy the problem very quickly before using up all my ammunition. To do this, I had to break out the ammunition belts, remove the excess ammo that had built up in the turret and reload my guns.

By the time I finished, I was ringing wet with sweat. Then my clothes started to freeze since the temperature was below zero. I had an awful time trying to keep warm. As we came off the target, we saw German Me-109s but they did not attack since we had P-38 fighters for escort.

Everyone breathed a sigh of relief on our return flight as friendly territory came into view. As we approached our airfield, crippled planes fired their red flares as the tower gave them priority to land. The ground support crews were standing by with fire trucks and ambulances. After landing, it was off to debriefing, then to the flight surgeon, then two ounces of whiskey and then to the mess hall. I drank the whiskey today.

We were all very tired, so we hit the sack early. When I got back from a mission like this, I would feel physically and mentally drained. Since our planes were not pressurized, we had to be on oxygen for long periods and then there was the mental stress. We had hit the factory and marshalling yards, though partially obscured by a smoke screen. We had flown at 21,000 feet and dropped fifty-eight tons of 500-pound bombs. Twelve of our aircraft suffered damage from flak. Flying time was seven hours. By now, I had twenty missions under my belt and thirty to go to complete my tour. I had survived.

May 31, 1944

As part of operation "Tidal Wave," our target this time for my twenty-second mission was to be the dreaded Ploesti oil refineries in Rumania in central Europe. "Tidal Wave" was the code name given for the destruction of the oil refineries that fueled the Third Reich's war machine. This was the group's third mission to this target. Our crew had not been scheduled for the previous two.

Encountering flak during our bombing raid of Ploesti oil refineries in Rumania.
Courtesy: John L. Lenburg Collection

We knew the Germans would heavily defend it, since Ploesti was the third most defended target in Europe. The Germans would use smoke generators to obscure the targets, so we would have to drop our bombs by using the Pathfinder method. This meant the lead plane was equipped with radar.

Our losses were light on this mission. We did not encounter any sizable number of enemy fighters, only heavy ack-ack. I did have one anti-aircraft shell blow up close to my turret. I heard the bang and then suddenly black smoke covered my turret. It scared the hell out of me. The force from the explosion caused my turret to move about six inches. I really checked myself over afterwards to see if I had been hit. I was lucky since our plane had been hit by some of the flak. I could not believe that I was still in one piece.

Twenty-seven out of thirty-two aircraft had flak damage. These small jagged pieces of flak can kill, wound or maim an individual. Two crewmen suffered wounds.

Thirty-two our aircraft dropped sixty-four tons of 500-pound bombs and the mission took seven-and-a-half hours. With the closing of the month of May, our losses had been fairly heavy in men and material.

In the twenty-six months that Ploesti came under attack, more than 13,469 tons of explosives and incendiary devices were dropped from 7,500 bombers. We lost more than 350 of our planes on those missions to Ploesti.

June 4, 1944

The mission was the marshalling yards at Turin, Italy. Our radio operator Ralph Wheeler was ill so Sid Rotz from Lieutenant Hanann's crew flew as a replacement. We encountered MC-202 Folgore Italian fighters and German Luftwaffe twin-engine JU-88s aircraft. Some of the JU-88s carried thirty-seven-millimeter cannons. They would sit out of our range and lob their shells into our formations. Thirty-eight of our aircraft hit the target with seventy-one tons of 500-pound bombs. Flak was encountered over the target area. Nineteen of our aircraft suffered damage.

June 5 and 7, 1944

On June 5, 1944, we bombed the marshalling yards in Faenza, Italy, and, two days later, on June 7, it was the harbor installations at La Spezia. With our primary target overcast, reducing our visibility, we hit our secondary target instead, the harbor installations at Leghorn. We had been flying quite a few missions in support of the Allied offensive in Italy. The offensive was going very well, and the Allies had just captured Rome. We flew at 22,000 feet and dropped eighty tons of 10,000-pound bombs. Fifteen aircraft were damaged from flak.

June 9, 1944

On our next mission, our orders were to hit the Allach Motor Works near Munich. We had to abort the mission since one of our engines developed an oil leak.

June 11, 1944

I was glad that we were not scheduled for our June 10 mission to Wiener Neustadt. Four aircraft were lost, two crash-landed and thirty-three out our returning airmen were injured. Our target for this day was the train ferry at Smederevo, Yugoslavia. Colonel Harrison led the attack units. We rendezvoused with the 485th Bomb Group between Altamura and Gravina at 0707 hours and then with the 55th Bomb wing at 01717 hours. Thirty-five of our aircraft dropped eighty-five tons of 500-pound bombs on our target with good results. We ran into intense anti-aircraft gunfire. Three of our aircraft were damaged and one crewman sustained a flesh wound. Over a hundred barges were observed on the Danube near the target.

June 13, 1944

The Milbertshofen Ordinance Works in Munich was our mission for this day. Seven-hundred-and-fifty heavy bombers with fighter escorts bombed targets in the Munich area. This was my first mission into Germany. We had received intelligence reports that German civilians had been hanging downed American airmen there. The German propaganda ministry had been inciting the population by calling us "Luft Gangsters" and "Terror Fliegers." We were not expecting to have an easy run since this time our mission was inside Germany.

That morning, before leaving on the mission, Mike Brown and I had eaten some locally grown cherries given to us. As we neared the target area, I started getting extremely severe intestinal cramps. Finally, I could not stand it any longer.

I told Erf that I had to come out of my position and go to the bathroom. I slid out of the turret, grabbed an oxygen bottle, and headed for the bomb bay. When I got there, the bomb-bay doors were already open, since we had started to make our bomb run on the target. Who do you think that I found already there? "Mike."

As Erf released the bombs, here was Mike and I relieving ourselves with flak busting all around us. (I guess we both gave Hitler a piece of our minds.) I had just settled back into my turret when I heard this over the intercom: "Ball turret gunner to pilot. I think that we have developed an oil leak because I'm getting little brown specks on my turret view plate." We never told Rube what had happened.

The group dropped sixty-six tons of 500-pound bombs on the target. We encountered intense anti-aircraft fire and some Me-109s. Our crew chief had some patchwork to do after this mission, although there were no injuries. It was a beautiful sight flying over the Alps with its snow-covered peaks.

After completing twenty-five missions, our crew was eligible to go to a rest camp on the Isle of Capri. We never got there, although we ferried several other crews to Naples for this purpose. In fact, while doing this mission, we received news over our radio that the Allies had landed at Normandy. Our crew's names never seemed to be on the list to go.

June 14, 1944

The next day our mission was to hit an oil refinery in Petfurdo, Hungary. We rendezvoused with our P-38 fighter escort over Banjakluka, located in the northwestern part of the country, at 1035 hours. They furnished air cover on penetration and withdrawal.

Our group dropped seventy-five tons of clustered and 250-pound bombs. We observed flak with red bursts and great many explosions and fires over the target area, with smoke rising to 15,000 feet. We observed a new airdrome under construction near the target area.

The mission took just a little more than six hours. We lost T/Sgt.

Top: A B-24 bomber goes down in a fiery blaze over Italy on the June 22, 1944 mission. *Bottom:* Carl Hobart, nose gunner, killed. *Courtesy: U.S. National Archives and Records Administration*

David Walsh of the 762nd Squadron that day. While throwing out bundles of chaff out of his plane, he fell out of the waist window. Two planes were also lost.

June 22, 1944

On June 22, it was a marshalling yard at Castel Maggiore, Italy. The plane on our left took a direct hit from a rocket that blew its nose turret off. Killed was nose gunner Carl Hobart. Then German Me-109s and Fw-190s hit us, even though we had fighter cover. Three Me-109s attacked our formation with rockets. Thirty-seven of our aircraft dropped seventy-four tons of 500-pound bombs on the target. Flak was heavy and intense; one plane was lost, seventeen planes damaged from flak and rocket fire and one failed to return. Two crewmen were wounded.

June 25, 1944

My thirty-second mission was an oil refinery in Sète, France in the southern region near the Spanish border. We picked up our fighter escort of P-38s near Toulon. We came under attack by thirty-five enemy fighters. Our fighter escorts, P-51 Mustangs, shot down three of them while we lost two of our planes. We hit the target with thirty-six tons of 500-pound bombs from 17,800 feet. Smoke from the exploding bombs rose to 16,800 ft. This mission was a long one: It lasted nine-and-a-half hours.

June 26, 1944

This mission was to be another big one. The 15th was to hit with our bombing wing the Florisdorf Oil Refinery in Vienna, Austria, while the rest of the 15th attacked other targets in the area. Our group formed into two attack units and rendezvoused with the 464th, 465th, and 485th groups. On this mission, I saw more air battles take place then I had seen on all our other missions put together. There were planes all over the sky. Our guns got a workout. I think that the Germans were getting desperate. Fighter escorts of P-38s and P-51s furnished penetration and target cover. We engaged enemy aircraft at 0946 hours near the target area. We had to be very careful so as not to hit any of our planes.

June 26, 1944, Vienna, Austria: Into the mouth of hell as our bombing group attacks the Florisdorf Oil Refinery. *Courtesy: John L. Lenburg Collection*

The Germans hit us hard with Messerschmitt Me-110s, Me-210s, and Me-410s (a modified version of the Me-210) firing their rockets and cannons. They also threw up a curtain of flak over the target area. We had casualties again on this mission, but we made it back.

Twenty-nine of our planes hit the target but twenty-five returned. Four parachutes descended from the first plane shot down, two parachutes from the second, and nine parachutes coming from the third. Flak damaged eight planes. Some of our planes came back looking like a sieve. I think we needed a rest, but our names were still not on the list for going to the rest camp on the Isle of Capri.

I was beginning to think that maybe I had a chance to complete my tour of fifty missions but, on the other hand, I felt that I was living on borrowed time. In some cases, we had lost our third and fourth replacement crews. It was getting to the point that we did not know the new crews that were flying with us on our missions.

I also noticed there was not as much joking or camaraderie between crews anymore. You did not really want to get too close to anyone. Now, I was having trouble sleeping at night as I was beginning to dream I would be shot down. I guess, almost all of us began to experience some problems as the missions piled up. Some pilots aborted their mission because of sick crewmen. Nerves and combat fatigue were most likely the two biggest reasons.

Flying one of these four-engine planes in combat was no easy task. I saw what our pilot and co-pilot went through to bring us back to the base safely. Many times, upon returning, Erf would be dripping with perspiration, even though we had been flying in sub-zero temperatures. Erf was the only pilot with whom I flew a completed mission.

On one occasion, they assigned me to another crew for a mission. As we were going down the runway to take off, I noticed that the pilot had not changed the pitch of the props too high. Luckily, I noticed this or we would have never made it off the ground. He had time to abort our takeoff.

To fly one of these planes in formation, with thirty-five to forty other planes or to bring it back with its nose, tail section missing and land safely, took great skill. It took lots of guts to fly through the flak, knowing that any one of those upcoming shells could be a direct hit.

If you were hit, you would be one of thousands of pieces falling to earth. In the event you were unfortunate enough to be shot down, it took even greater courage and skill to stay with the aircraft so that the other crew members had a chance to bail out.

June 29, 1944

On the evening of the June 28, 1944, an officer posted our crew's names on the list of crews that would be flying the next day's mission. Checking the list outside the orderly room to see which crews would be flying the next day was always an evening ritual.

We awakened at 0430 hours to start getting things ready for the mission that day. At the briefing, many of us gazed up at the large map of Europe, outlining the mission in blue yarn, which showed the route and target for that day's mission. It was a synthetic oil refinery, the Blechhammer South Oil Refinery, in Germany. This would be our deepest penetration into Germany and would be a long and dangerous mission. We would be dropping 500-pound bombs. Since this mission was to be about 2,200 miles round trip, our fuel consumption could be a problem. We carried 2,700 gallons of gasoline and consumed it at about a gallon a mile.

To reach our target, we would be flying up the Adriatic over Yugoslavia, Hungary, and then over Czechoslovakia into northern Germany. This route was called "flak alley," since the Jerries had such a heavy concentration of anti-aircraft guns. They could also put up 700 to 1,000 fighter planes.

We moved up to the flight line to get things ready for the mission, but after waiting several hours, they scrubbed our mission, much to our relief, because of bad weather over the target area. I hoped that maybe they would pick us, or if they rescheduled the mission the next day, it would be a different target. That did not happen.

June 30, 1944: Blechhammer in view with bad weather and poor visibility developing. *Courtesy: U.S. National Archives and Records Administration*

BLECHHAMMER

June 30, 1944

It was Friday and they scheduled our crew to fly after scrubbing Thursday's mission. Sergeant Cutler awakened us at 0430 hours in the morning. At the briefing, we learned that the mission to Blechhammer was still on. Our group was to lead a force of 500 to 600 planes of the 15th Air Force on this mission. We were to fly number two position in the high box of the lead attack unit and Lt. R.G. Evans was to be to our left in number 3 position. Capt. John "Ham" White was leading our box. Colonel Harrison was to lead our group in the lead box of the first attack unit. After our pre-flight inspections, everyone boarded the aircraft to prepare for take-off.

At 0625 hours, thirty-six aircraft took off. Our group rendezvoused with the 485th Bomb Group between Altamura and Gravina. Then, at

Messerschmitt Me-109 fighter planes being assembled for action in a German assembly plant. *Courtesy: German Federal Archives*

Side mounted remote gun turret of German Messerschmitt Me-210 fired during air combat. *Courtesy: German Federal Archives*

approximately 0725, we rendezvoused with the rest of the 55th Bomb Wing over Spinazzola, with the rest of the 15th following. We were to pick up P-38s as escorts but because of bad weather, this never happened.

I was in my usual position, in the top mount of the plane, when we approached the most southern part of Hungary. We started running into the bad weather. There was a high dense overcast with high clouds. This condition cut our visibility down to almost zero. As a result, they aborted the mission. In turning back, we lost visual contact with the other planes in our attack unit but were able to keep visual contact with the planes of our box.

As we came out of the clouds, I noticed that our box had separated from the rest of our group. Our box was flying alone, and I did not see the rest of the planes in our group. I did not know what happened to them since I heard nothing come over the intercom.

My turret was facing forward, and I noticed a large formation of planes several miles in front of us. We seemed headed for them. Just as I started to swing my turret around to the right, shells began exploding around us. I thought that it was coming from anti-aircraft guns on the ground. That was the last thing that I remember. We have theorized afterwards what must have happened. My turret had taken a direct hit from a 20-millimeter cannon shell, knocking me out. Thirty-five to forty German Me-210 fighter planes, coming in waves of four and raking our planes with 20-millimeter cannon fire, had attacked our box. They began firing as they were coming through the overcast sky.

The next thing I recall, I was having some difficulty breathing. Heat from the fire on the flight deck sucked up into my turret. It was then I realized that something had happened, but I was not yet sure what. I had been wearing sunglasses and all I had left were the frames minus the glass. My flak suit was shredded. Still, I do not think I realized how badly we had been hit; I also do not know how long that I had been unconscious. I recall seeing a German fighter on our tail, his guns blazing at our rear. I tried to operate my turret, so I could bring my guns around to fire at him. The turret was inoperable. It was probably at this point I fully began to realize my turret had taken a hit.

I disengaged the turret from the plane's electrical system, so I could operate it manually. I cranked it around, so I could fire my guns directly into the fighter who was still on our tail. There was no response from the tail gunner's position. My tracers were going right into him. He started emitting some smoke and then peeled off. At this point, the heat from the fire on the flight deck was becoming unbearable. I was having a very difficult time breathing from the heat sucked up into the turret. The fire on the flight deck worsened. Suddenly, I had a sick feeling in the bottom of my stomach.

Top: A B-24 Liberator is hit by enemy fire during the raid on the oil refinery at Blechhammer. *Bottom:* Fourteen days earlier, on the June 16, 1944 mission to bomb the synthetic oil refineries, a B-24 bursts into flames after being hit by a German Me-109 fighter plane over Vienna, Austria. *Courtesy: U.S. National Archives and Records Administration*

At that moment, I knew we were in serious trouble. We were going down.

The following quotation was taken from the 460th missions report to the 55th Bomb Wing Headquarters concerning the June 30th mission to Blechhammer:

> *At 1045 hours 45-50 Me-210s were observed in the Lake Balaton area. 35-40 a/c [aircraft] attacked, concentrating their fire on the high box of the first attack unit. They came in from 7 o'clock high in eight waves of four abreast and closed most aggressively to within 50 yards. Seven other Me-210s attacked singularly from 6 o'clock and 11 o'clock level. There were two encounters with Me-109s. In one case, a pilot emerging from a cloud found a Me-110 attacking a B-24, which had one-gun operative, altered his course, interrupted the fighter and shot him down. Three a/c [aircraft] destroyed were claimed. From the high box of the first attack unit, 4 a/c [aircraft] are missing, though only one was observed to go down, twenty miles N. of Lake Balaton, with #3 engine and bomb bay in flames. No chutes were seen.*
>
> *(EEA Reports filed). The surviving a/c [aircraft] of this box evaded e/a [enemy aircraft] by seeking cloud cover, although it was damaged considerably by e/a [enemy aircraft] fire, returned to base where a crash landing was made successfully. One man sustained a slight scalp wound.*

Top frame: June 30, 1944: Our B-24 bomber is shot down. *Middle frame:* "Miss Fortune" descending fast. *Bottom frame:* It explodes upon impact over Lake Balaton. *Courtesy: Nándor Mohos*

GOING DOWN

We still had a pretty good load of gasoline and a full load of 500-pound bombs. I had seen other planes in this situation blow up, with no one getting out. It was at this point that I felt I was going to die and asked "God" to save me. I reached under the seat and pulled the seat release, which allowed me to drop out of the turret to the flight deck. I then realized how badly damaged our plane was and the extent of the raging fire on the flight deck. Erf looked around and motioned for me to get out. As I looked at him, the thought crossed my mind if I would ever see him again.

Erf had told Alan to ring the bailout bell, but the noise was so deafening that I could not hear it. He and Alan were trying to keep the plane in a level flight and avoid going into a spin. If the plane started going into a spin, none of us would be able to get out. The centrifugal force of our spinning B-24 Liberator bomber is all that held us. It was just like an ocean ship going down. Erf was the captain and stayed at the controls until the end to make sure everyone had a chance to bail out.

I pulled the release to my flak suit; it fell off. At the same time, I pulled off my flak helmet. The fire was on the flight deck by the gas gauges on the rear bulkhead. I usually stashed my parachute in this area. I reached for my chute pack, snapped it on and headed for the bomb bay. In doing so, I burned my right hand reaching into the fire to retrieve my chute. Then the back of my jacket caught on fire from the intense heat. The forward bomb-bay doors were open, and four bombs were missing.

Mike Brown was already on the catwalk in a crouched position in front of me, but he seemed to be hesitating to jump out. My feet were right on his shoulders. We still had a partial load of bombs and I could still envision the plane blowing up since it was burning so badly. By now, my jacket was burning pretty good as I was waiting for Mike to jump. I guess at that point, I gave Mike a little nudge. Mike went out and I right after him. I guess that helped him get out. As soon as I hit the slipstream, my jacket fire quickly extinguished. I did not open my parachute right away since I thought we were at a higher altitude, not realizing we had dropped down to about 12,000 feet. I started counting slowly in order to delay opening my parachute. I think I got up to about fifteen when I decided to pull the ripcord. It was a good thing that I did not delay my parachute opening any longer, for I had not realized our plane had dropped down to a lower altitude. If you bail out above 10,000 to 12,000 feet, you should delay pulling your ripcord because of the lack of oxygen in the atmosphere.

I was tumbling slowly through the air as I was falling down to earth. When I reached for the ripcord handle with my right hand, I found it was not there. In my haste, I had snapped the parachute on backwards and the handle was on the left side instead of the right. Luckily, I found it and pulled it with my left hand. By then I was falling headfirst, and it was a welcome sight when the chute flew past my face and opened up. When it did, I suddenly jerked into an upright position and my descent slowed considerably.

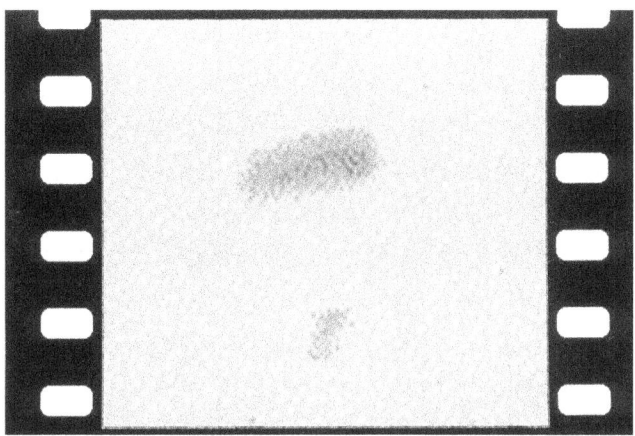

An American airman from our bomber group parachutes to earth before his B-24 went down. *Courtesy: Nándor Mohos*

After leaving the deafening noise in the plane, suddenly everything was quiet. As I was floating down, I began to realize the full extent of my injuries. I burned my forehead and right hand; my right jaw had split open. I had pieces of 20-millimeter in my face that were about the size of shotgun pellets. A Hungarian doctor who examined me later told me this. He kept pointing at them and saying, "husz millimeter" (Hungarian for twenty). The largest piece was stuck in the corner of my left eye. Luckily, for me, it must have been pretty well spent when it hit me, since it was stuck in the right corner of my left eye by my nose. Another piece went through the left corner of my mouth and another through the right corner.

The following is an account of what happened written by our co-pilot Alan Barrowcliff:

> *Approaching Hungary, we encountered many large, white, billowy clouds. The visibility became so bad that it became very difficult to keep the rest of the boxes of our group in sight. At 17,500 feet near Lake Balaton, Hungary, we were attacked by twin-engine Me-410s. They stayed just above the clouds and shot at us with 20-millimeter cannons, but we could not see them or reach them*

effectively with our 50-caliber machine guns. Later, when they started to make passes over us from the rear and going forward, I saw tracers from Jack Nagel's nose turret hitting the port engine of one of the fighters and putting the engine on fire.

One of the engines on our left wing caught fire and brown smoke started to come up from under the instrument panel. The glass tubes of the gas gauges at the rear of the flight deck had been hit and were burning. It all sounded like hail on a tin roof when we were getting hit.

Strings of 50-caliber ammunition were hanging down onto the flight deck from John Lenburg's upper turret, then the fire started exploding the rounds of ammunition, adding to the noise.

I don't recall if we had radio communications with the rear of the plane; however, due to the conditions, we had on the flight deck, Erf and I decided we were in bad shape and had better bail out. He told me to call over the intercom and to ring the bailout bell.

By this time, we had dived down to 12,000 feet to try to escape the German fighter planes and leveled off to let everyone jump. While Erf held the plane level, I pulled the handle, which was supposed to open the bomb-bay doors in an emergency. On the second pull, all of the bombs should be released. The forward port door was the only one that opened and released the four 500-pound bombs in that bay.

John Lenburg had gotten out sometime before I tried. The flames from the gas gauges singed me on my first attempt to get off the flight deck through the small doorway leading to the bomb bay. Then I opened the top hatch, which was behind the pilot's seat and in front of the top turret.

After climbing up and looking out, it didn't seem that I would be able to jump from there without hitting the propellers of the two inboard engines or, if I got past them, I would probably hit the rudders and knock myself out or get killed instantly. From this

Twin-engine German Me-410 that attacked "Miss Fortune" and our crew.

spot, I could see the flames coming out of the back of the plane and that the left wing was also on fire.

So, I dropped back onto the flight deck. I felt something hit my ankle. Seeing a small rip on my flying boot and later a nick on my G.I. shoes, I figured that it must have been a bullet from the exploding rounds on the flight deck.

I realized then that there was no alternative and that I would have to dive through the flames to get to the doorway, which led to the bomb bay.

As I dove through the flames, I saw something burning on my right shoulder and figured that it was my parachute strap that had caught fire. The catwalk of the bomb bay was three or four feet lower than the flight deck level; so, I had to reach down and pull myself through the doorway and then twist my body to the right in order to get through the bomb bay opening.

By this time, I didn't think that I had a parachute anymore on my back since I had seen something burning on my shoulder. But after being partially burned on my hands and face, I knew that I didn't want to be burned to death and would jump without a parachute.

The next thing that I was conscious of was a sharp pain in my right groin area. A parachute strap pulling up tight caused the pain when the chute opened. It was a beautiful sight to look up and see a white canopy all in one piece, fully opened, and not on fire. What had burned on my shoulder was the first-aid kit that was tied to the straps.

I have no recollection of counting to ten before pulling the ripcord to open the parachute, and I must have dropped the metal D-ring right away. The plane was not very far above me, so I must not have waited too long before pulling the ripcord. I don't remember actually leaving the plane; I could have passed out from the pain in my back while twisting to pull myself out. My watch showed 10:00 a.m.,

When I looked down and around, I was floating right over the middle of Lake Balaton. However, the wind was blowing me towards the north shore. There was no sensation of falling at that height. After all, of the noise of the flames, exploding ammunition, and cannon shells hitting the plane, it was relatively quiet now.

Facing towards the north, I could see four chutes in a row below me in the distance to my right; above me, on my left, I could see one more, which had to be Erfeldt's since he would have been the last one to jump. Later, I found that the four chutes belonged to Nagle, Hendricks, Brown and Lenburg; Bernhardt had jumped earlier. These were the only men to survive; Wheeler, Waits, and Troy died in the plane.

There were several burning wrecks on the ground and our plane made a slow turn to the right. It then crashed into a hillside many miles away. The weight of the bombs still on the right side of the plane probably made it turn in that direction. A German Me-109 made a pass close to me below my feet but continued towards the west end of the lake.

Bernhardt gave this account of what happened in the rear of the plane. He and Waits had traded gun positions for this mission, so he had been in the ball turret and Waits in the tail turret.

On the way to the target, our box of aircraft was separated from the rest of the formation by bad weather. We were attacked by Me-110s (I saw 12, but there may have been more, as visibility was poor). Four of our aircraft were shot down out of the five. I had been wounded, the interphone had been shot out, tail gunner got one Me-110, and the No. 2 engine caught on fire, in that order. The entire section of wing caught fire, and then I saw fire in the bomb bay. I believe the bomb load was salvoed.

I ripped off my flak suit and went back to get the tail gunner out. On my way I tried to open the camera hatch, but it would not open. It had been in good condition at time of takeoff, but I think it had been welded shut by heat or jammed by fragments. When I got back to the rear, I found the gunner dead and the turret damaged by gunfire. Then I turned back in the tail of the plane, but the flames were right in my face.

I fought my way back to the waist and got my parachute, which had been wrapped in a flak suit. The other waist gunner was standing there in the flames, seemingly paralyzed by fear, wearing a flak suit. I motioned to him to put his chute on but got no response. Then I started out the waist window, but got stuck, being held partly by centrifugal force. The other waist gunner then pushed me out. I was semi-conscious, but believe I opened my chute at once. I did not see his aircraft after getting out.

The following is an account of the events given by radio operator

Marvin Wycoff on Lieutenant Sorgenfrei's crew, a new replacement crew, in the 762nd Bomb Squadron. They were flying in the high box of the second attack unit until they encountered the bad weather. Separated from the planes in their box, they saw our planes and attached themselves to our box:

> *We continued to fly missions at a rapid pace, skipping the rest period at Capri, in the hopes of finishing our 50 missions and return to the States. The road to this objective proved rough, with many obstacles along the way. One of these obstacles occurred on the trip to Blechhammer. If you were on this mission, you will remember the bad weather over Hungary. Our formation entered the clouds placing us in extreme danger of mid-air collisions. I can't remember where we were in the overall bomber formation, but Ken [Sorgenfrei] applied more power and climbed as rapidly as possible with our bomb load still aboard. We cleared the clouds and found three other 460th bombers in a small formation. We attached to them as another B-24 approached which I think was to our left-wing position, now making a box of five B-24s. There were no other 460th planes in sight at our higher altitude, probably near 20,000 feet.*
>
> *A long way off we could see a large formation, near our altitude, and even though I don't think our planes were talking by radio, our crew thought we were drifting closer to join this formation and perhaps drop our bombs on their target. As these planes got closer, however, they were not B-24s nor were they B-17s. They were a large formation of Me-210s; at least forty and they lost no time initiating a vicious attack. These fighters came in waves of four abreast from the rear with another wave following a short distance behind. They would open fire as soon as the first wave broke off their attack at a very close range. The plane beside us was hit hardest on these early waves and was soon burning and going down. This left us "Tail-end-Charlie" and now we were receiving the most attention from the Me-210s. We were taking a*

beating from their guns, which were not machine guns but were 20-millimeter cannons. Other fighters, whom I couldn't see from the waist position, came in from the front at the same time but the heaviest and the steadiest attack came from the rear.

We had taken hits on the trailing edge of our wing, just behind the main landing gear, near engines one and two. There were large pieces of sheet metal in the flap areas just waving in the wind? We were soon running low on ammunition, as these Me-210s never let our guns rest. Ken yelled something like "Hold on! We're going down to the clouds" and about two seconds later, I was practically plastered to the roof of the plane. It took Ken and Ray Swedzinski to level our plane off in the clouds while still carrying a full bomb load. Here we

15th Air Force B-24 Liberators fly through flak during a bombing raid of Ploesti oil fields in Romania in April-November 1944. *Courtesy: U.S. National Archives and Records Administration*

> *played a deadly game of hide-and-seek for the next few minutes with the fighters going in out of the clouds. They wouldn't come into the clouds with us and we found ourselves once again, in a badly damaged plane, trying to get back to Spinazzola, some two hours or more away.*
>
> *We finally limped in. All the rest of the group had landed, and we were able to make a successful landing, even though the group called it a crash landing. During debriefing back at the base, we kept watching for some of the planes to return that were in the high box. None returned. We were the only surviving plane of the five.*

As I floated back to earth, a German Me-109 fighter buzzed me. I was praying that he would not try to cut my legs off with his prop as some Jap flyers had done in the Pacific. He came close; in fact, too close. After several passes, he waggled his wings and then banked to the right and flew off. Most likely, this same plane buzzed Alan.

The same kind of an incident happened to Bernhardt as he recounts in his version of what happened:

> *On the way down, I was buzzed twice by a Me-110. Each time the aircraft came straight at me, I dodged it by spilling air from my parachute. (I was told later by some Hungarian pilots that there had been several such cases, where a pilot clipped shroud lines on a parachute as a form of sport.) I landed in a wheat field at the northwest corner of Lake Balaton.*

I suddenly realized I had a new problem. If I continued descending as I was, in all probability, I would land in the water. I knew that I had bailed out over southern Hungary but did not realize that we were over Lake Balaton. I had my Mae West on that I usually wore under my parachute harness. It can be quite a trick getting out of a parachute harness along with a wet parachute in the water. With a wet chute and its shrouds on top of me besides heavy wet clothes, there was a good chance that I would drown before getting out of the parachute and inflating my Mae West. I kept spilling the air out of my chute by yanking on one of the shroud lines. By doing this, it helped change the direction of my descent so that the slight wind started carrying my chute more towards the shore. I will tell you, it was pretty scary doing that, since I started to drop straight down and wondered if the chute would fill with air again. It looked like I was finally going to make land.

The German propaganda machine had been hard at work inciting the population against any American airmen with whom they might come in contact. The following is a statement on the propaganda taken from Bernhardt's statement to Allied authorities on March 1945:

> *Germans spread propaganda that American bombers were dropping explosive fountain pens, pencils, dolls, toys, etc., over Hungarian towns. Hungarians reported that this caused beating and killing of American airmen. Hungarians were told, in newspapers and posters, that it was their duty to kill any American airmen they saw coming down in a parachute, as these men were coming down only to kill Hungarians. This held true even if men were bailing out of a burning aircraft, etc. Many men who landed safely were in the hospital due to injuries inflicted by Hungarians.*

POW

As I got closer to earth, another obstacle appeared. I was going to land in a vineyard on the side of a hill. The vineyard was full of poles stuck in an upright position supporting the grapevines. I was going to have to try to negotiate myself between them. If I could not, then I would be in for a real problem. I did manage to slip in between the rows of poles. As I hit the ground, I did a front roll between my legs. (They taught us to do this as part of our training.) In doing so, I hit my chin on my knees and chipped all of my front teeth.

As I slowly began to rise and try to regain my senses, two Hungarian soldiers ran up to me and with their rifles pointed at me, hollering, "Pistola! Pistola!"

I indicated to them that I did not have a gun.

As the soldiers started helping me gather up my parachute, some of

Gary Aerial Gunner Lost During Raid Over Austria

Tech. Sergt. John L. Lenburg, age 20, graduate of Lew Wallace high school, is reported missing in action over Austria since June 30, according to a war department telegram received by his uncle, John G. Lenburg, 3908 Jefferson, with whom he made his home.

Meanwhile, in the Calumet area, one soldier was reported killed in France, another missing over Germany, and two wounded in action in Italy.

Notice also has been received by the family of Pvt. Andrew A. Pazak, age 27, who was wounded in action May 25 in Italy, that it had been found necessary to amputate his legs. Pazak wrote his brother, Mike Pazak, 1720 Van Buren, with whom he made his home, that he had been shot in the legs. He had been overseas a year and went through the campaign in Sicily.

Sergt. Lenburg was a gunner-engineer on a Liberator bomber and according to a recent letter from him had been on 31 missions up to two weeks ago. He had received the air medal and several oak leaf clusters. He had been in service since December, 1942, his high school diploma having been issued since he was inducted. He went overseas in January. He is the son of the late Mr. and Mrs. Leo Lenburg.

Pfc. Alfred J. Van Holsbeck, age 21, was killed in action June 6 in France, according to a notice received by his parents, Mr. and Mrs. A. A. Van Holsbeck, 1420 119th, Whiting He attended Whiting high school and worked for his father in the upholstery business before his induction in January, 1942.

Staff Sergt. Robert Holly, age 22, a bomber gunner, has been missing in action since June 20, his parents, Mr. and Mrs. Byron Holly of Schneider have been informed. He is a graduate of Lowell high school in the class of 1941. He enlisted in the army air forces in January, 1943.

Mr. and Mrs. Ernest Hodges of Crown Point have been notified that their son, Sergt. Milton Hodges, was wounded in action and has been awarded the purple heart. He was with the 5th army in Italy. A letter from him disclosed that he is in a hospital and recovering. Pfc. Donald E. Cyborski, son of Mrs. Pearl G. Gorman, 229 Ogden, Hammond, also is reported to have been wounded in action in Italy

Newspaper account back home reporting me missing in action after the Blechhammer raid. *Courtesy: John L. Lenburg Collection*

the area farmers came running over and started gathering around me. They had clubs, rakes, pitchforks and scythes. They began shouting at me while shaking their fists and waving their clubs. One of the peasants had an old blunderbuss gun. This type of gun had a gun barrel that looks like a funnel on the end.

Some of them tried to spit on me and then hit me with their sticks. Then all of a sudden, the one farmer with a pitchfork made a lunge at me. One of the Hungarian soldiers threw the butt of his rifle up and deflected the end of the pitchfork. If the soldier had not have done this, I probably would not be writing about this episode.

The other Hungarian soldier started scolding the people. They backed off. However, they kept following us, shouting and shaking their clubs and fists at me all the way down through the vineyard to a road. Here the soldiers put me in a Ford station wagon. At this point I think that I starting to go into shock. I started shaking and felt like I was going to pass out. My right jaw that had split open was killing me. My face was bleeding and hurting from the facial wounds and burns. I could hardly open my mouth.

The following is an account of what happened to our crewman "Pappy" Bernhardt when he landed:

> *As soon as I landed by parachute in a seriously burned condition, I was attacked by peasants carrying pitchforks and spades and almost beaten to death. I was saved only by the Hungarian military.*

Hungarian soldiers drove me to a small village where I was brought before several German officers seated at a large table in what appeared to be the town square. One of the officers asked me my name and with what outfit I was. I told him my name, rank and serial number. All that I was required to give under the "Geneva Conventions Agreement."[3] They did not press the issue but decided to take my clothing away from me. So, I lost my coveralls, slightly burned jacket and G.I. shoes. In its place, they gave me some parts of a heated flying suit and felt shoes to wear with it that were too big for me. Why they did this, I really do not know. One of the German officers said, "For you, the war is over."

He was wrong; the toughest part was just beginning.

They kept me standing for a while until an open bed truck pulled into the square, and then told me to get into the back of it. Some of our crew members, Erf, Brown, and Bernhardt, along with a couple of other captured airmen were on the truck. It had been traveling around the countryside picking up captured airmen as several other planes had been shot down.

German fighters had been attacking our box by coming in four abreast

---

[3] *The Geneva Convention was a series of international agreements signed by sixteen nations on the humane treatment of sick and wounded prisoners of war of the armed forces in wartime. Germany was one of the signers.*

firing 20-millimeter cannons in our wake. As a result of the attack, four aircraft from the 460th, Captain White's, Lieutenant Erfeldt's, Lieutenant Evans's of the 760th Squadron and Lieutenant Champlin's of the 762nd Squadron were shot down. Lieutenants Champlin and Sorgenfrei were in the 762nd Squadron. They had tried to attach their aircraft to our box after separating from their own boxes due to the bad weather. Lieutenant Sorgenfrei was the only one able to get away from the attacking German aircraft. This had been a very costly mission in men and planes. Killed were seventeen airmen and twenty-four became POWs.

Bernhardt, known as "Pappy," severely burned his face and his head. He had no hair, eyelids or ears. The rest of his head was just a black mass of burned flesh. At first, I did not recognize him. I could see that he was in great pain, for he asked his captors to kill him.

Pappy said that Troy and Waits were killed. He also said Wheeler had sustained serious burns and he had helped him to get out of our burning plane, but Wheeler's chute caught on fire in the process. Since fire had

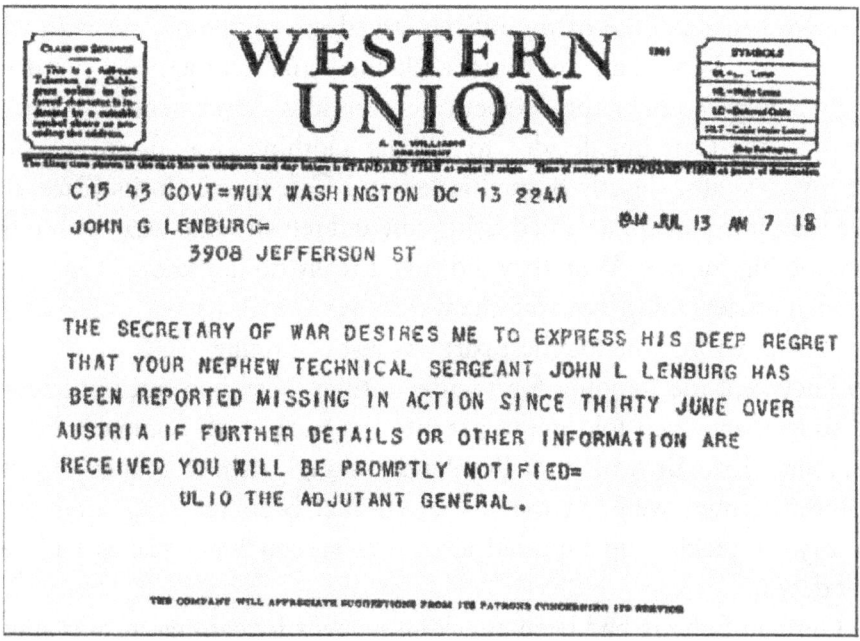

Western Union telegram sent to my uncle John notifying him that I was missing in action. *Courtesy: John L. Lenburg Collection*

damaged his chute, Wheeler's chute did not open, and he fell to his death. Later, they found his body north of the town of Badacsonytomaj.

They took us to an airfield where a German pilot jumped up on the backend of our truck. He kept saying that he was the one who shot us down, but Erf kept telling him that ack-ack shot us down, not a fighter plane. The pilot got mad at Erf, started shouting and jumped from the truck. Eventually, they took us to the small town of Tapolca and put us into a small hospital run by Catholic nuns. The nuns wore large, tri-cornered white hats. There were several other captured airmen already there. They treated my burns and facial wounds, putting me in a small ward with Erf, Brown, and others where I spent the night.

There was the possibility that our crew shot down or badly damaged at least three German fighters in this encounter. Barrowcliff reported that Jack, our nose gunner, was pouring rounds into one of the attacking enemy planes. He said that he could see pieces of the plane flying off and black smoke pouring out of one of its engines, causing the pilot to break off his attack. Pappy reported that Waits downed one Me-110 before he died in action. Then there was the fighter plane that I fired into us just before I bailed out. To confirm a kill, you have to observe it hitting the ground. This we could not do.

On Saturday, July 1, 1944, armed Hungarian soldiers escorted us to the train station, put us on a passenger train, and shipped us to Budapest. Along the way, we picked up Barrowcliff and some other captured American airmen. Bernhardt was in too bad a shape and did not make the trip with us. The train ran along the western shore of Lake Balaton. I noticed some of the passengers carried briefcases with them. When it was lunchtime, I found out what the briefcases contained: some bread and a piece of cheese or sausage—in other words, their lunch. The countryside was very picturesque, with the neat little cottages and white picket fences along the way. This was quite a contrast to the drab tufa blockhouses of southern Italy.

Royal Hungarian Hospital No. 11 (since renamed Tapolca Hospital) in Budapest, pictured in 1996, where I was treated for my injuries after my capture as a POW.
Courtesy: Nándor Mohos

ROYAL HUNGARIAN HOSPITAL NO. 11

We arrived in Budapest in the early evening at the bombed-out train station. It had suffered some damage due to the Allied air raids. There was a glass-domed skylight over where the trains pulled into the station. Most of it was missing. Some of the metal structure that held the skylight had become twisted and fallen. The 15th Air Force bombed Budapest by the day of our arrival.

Some confusion developed over what to do with us, until they took us to the Budapest City jail. There, they put us in cells and put me in a cell with a Hungarian national. The cell had two beds illuminated by one small light bulb hanging down from the center of the cell they left on at night. My cellmate was able to speak a little English. He told me that the RAF was going to bomb Budapest that night. Sure enough, the air-raid sirens went off with their loud wailing sound. This sound bothers me even today.

(The two towns next to Hobart, where I live, have volunteer fire departments. Therefore, the town uses these large sirens to summon their firemen when there is a fire. When I hear these go off, it sends chills up and down my back and I find myself listening for the drone of planes.)

The thunder of the exploding bombs and the anti-aircraft guns firing kept me up most of the night. The RAF conducted their air raids a little different than the U.S. Air Force. They would send a Pathfinder plane. Usually, a Halifax or Lancaster bomber into the target area that would drop slow, descending flares over the target. These would illuminate the whole area. Then the bombers would follow singularly and drop their bombs. How this Hungarian cellmate knew of the forthcoming RAF air raid has always been a mystery to me.

The next day, Sunday, July 2, 1944, they transported us to the Royal Hungarian Honvedspital[4] No.11 in Budapest. They delayed taking us there because of another bombing raid by the 15th Air Force hitting targets in Budapest again and the Allies doing 'round-the-clock bombing. It rained bombs most of the day. The Americans hit them in the daytime and the British at night. It was ironic that the 460th took part in this mission and here we were on the ground. My roommate and I spent more time under the bed again while the waves of planes passed overhead dropping their bombs. I can assure you that being on the ground during an air raid was a very terrifying experience. When the bombs dropped, they sounded like on-coming freight trains or a loud whooshing sound. They usually sounded like they were coming right down on top of you. It was late afternoon after the all-clear siren had sounded, and then the Hungarian authorities moved us to a hospital.

The following is an account taken from news dispatches covering the action by the 15th Air Force for July 2nd:

> *From 500 to 750 flying Fortresses and Liberators, including units from the 8th Air Force, which flew from England in a shuttle flight,*

---

[4] "Honvedspital" in Hungarian means "hospital."

ROYAL HUNGARIAN HOSPITAL NO. 11

bombed Nazi oil storage, refineries, and transportation facilities in Romania, Hungary, and Yugoslavia today for the second straight day. Dispatches quoted returning airmen as saying that the opposition was limited and all targets were plastered.

German news accounts reported:

Attacks were made for the second straight day in the Budapest area resulting in fierce air battles being fought. Thirty-five American planes were downed.

We arrived at the hospital in the late afternoon. As we entered the hospital, it was obvious that everything was in a state of confusion. Everyone seemed to be running around. Many of the bomb victims were arriving, resulting in chaos. We encountered some of the civilians; they tried to spit on us and shook their fists. They took us into a large room, which appeared to be Emergency Room, where a doctor checked us. After undergoing several days of the Allies' day and night bombing, they were not too happy with us being there. It seemed like they thought we were some of the flyers shot down over Budapest. Mike Brown kept trying to make them understand that was not the case. Since none of us could speak Hungarian, they did not understand. I really do not think that it would have made any difference since we were the enemy.

The doctor had us take our shirts off as he prepared to give us each a shot. When I saw where he was going to give it, I thought that this was the end. In fact, I think all of us thought that. Mike Brown loudly objected but the doctor gave him the shot in the chest just above the heart. It turned out to be a tetanus shot. He looked at my facial burns and wounds. He took a swab of cotton with alcohol and stuck it into my face. At that point, I think that I let out a yell, but he just smiled. It took quite a while before the piercing pain subsided. Later, a nurse put some ointment on some of the burns and wounds. She took tweezers and removed some of the 20-millimeter fragments from my face, with some left that were in deeper. Through the years, almost all of the pieces have worked their way out. I still have a few very small pieces in my neck and one in my finger on my right hand.

The following is an account of this incident given by Barrowcliff:

> *On Sunday, July 2, they took us to a hospital. They put us in the Emergency Room to get treatment. Injured civilians were also brought into that room due to the American air raids that morning. The civilians thought that we had been shot down that day and were extremely angry seeing us in there also getting medical treatment. They attempted to hit us and spit on us. They had to be restrained by the doctors and nurses.*
>
> *When a doctor started to take care of me, he produced a syringe to give me a shot. He pinched the skin together with his fingers on the upper left side of my chest. Never having had a shot in this area so close to the heart, I thought that maybe this must be the easiest way to put me out of my misery for good. However, I didn't feel faint nor have any dizziness, so I figured it must have been something for my burns. He also gave shots to the others. I found out later this was a tetanus shot.*

After they were finished with us, they moved us to a small ward. In the ward, there was a nurse in charge, named Elizabeth. She was not too happy about having to take care of these "Luft Gangsters." I guess she believed Goebbels's propaganda. Later, we moved to a large ward with other injured American flyers that had been shot down. Quite a few of them were in the hospital from peasants injuring them with their pitchforks, scythes and/or from shooting them. The ward was located on the top floor of one of the hospital's wings.

The following is an account given by Barrowcliff:

> *We were taken to a small ward of five or six beds. There, we had an English-speaking nurse. Even though she was an educated person, she believed propaganda, which said our bombers,*

A pair of German Bf-110 fighter planes fly over Budapest, early 1944 *Courtesy: German Federal Archive*

dropped dolls in little baby carriages that contained bombs to kill their children and we also dropped fountain pens, which had bombs inside of them. The power of propaganda was really strong to give her these thoughts even though she had never actually seen any of these devices. Her name was Elsbeth, which she said was Elizabeth in our language. She was in her late twenties.

Later we were taken to a large ward containing a lot of Americans who had been at the hospital for some time. The room contained double-decker cots with straw mattresses. Erf was in there with some of the others that were shot down on Friday, June 30. The majority of the patients in the room were there because of injuries they had received after being shot down.

The peasants in the fields had gotten to them and stabbed them with pitchforks, sliced them with scythes, and had even shot them with rifles and shotguns, which afterwards I realized how fortunate I had been to be picked up by soldiers. There must have been about twenty double beds around the walls and a few single cots in the middle of the room. At the other end of the room was a short Jewish gunner, Fred Meisel, who had his little finger sliced lengthwise with a scythe and some other injuries also. The peasants wanted to kill him because they thought that he was a Jew, but he had an American Indian-head nickel on his dog tag chain, which had a strong resemblance to his face. He convinced them that he was an American Indian and that was the only reason he was still alive.

Out in the fields of the countryside, there were old men stationed at every bridge-covered stream, even if the water was only a few feet in width. They were armed with small caliber rifles and shotguns. It was the women in the fields that had the harvesting tools.

Those of our crew in the ward were Erf, Barrowcliff, Brown and I. Erf and Barrowcliff had facial and hand burns. Brown had burns and a wound in the lower fleshy part of one of his legs from a 30-caliber bullet. I had facial and small hand burns, pieces of a 20-millimeter cannon shell still in my face, and a split open right cheek. The latter made it a very difficult to open my mouth. By the time I left the hospital, I could open my mouth reasonably well since much of the swelling had gone down.

Several days later, our navigator Matthew Hendricks joined us. He had a 20-millimeter cannon shell lodged against his spine. On July 8, 1944, "Pappy" Bernhardt was brought in looking like a mummy with his head all bandaged up. They put "Pappy" in a cot at the end of my bed. The pain must have been terrific since the burns covered his entire head.

Since "Pappy" was Jewish, they seemed to neglect him. The odor from the unchanged bandages was sometimes overwhelming. He told me that he wished that they would put him out of his misery. We all tried to help him take nourishment since the nurses seemed to neglect him. The only way he could eat was through a straw.

Jack Nagle was the only surviving crew member who came through the ordeal unscathed. They sent him to a prison in Budapest and then to the POW prison camp, Stalag Luft IV, in northern Germany. He spent part of his stay in the Budapest prison in solitary confinement. The only food he was allowed to have was bread and water twice a day.

Transport bus like the one the Germans used to move captured American and Allied servicemen to prison of war camps.

SENT TO PRISON

On Tuesday, July 11, 1944, we were informed the beds were needed by the hospital for more injured prisoners and we were going to be sent to another facility to complete our rehabilitation. This turned out to be a big joke on us. They loaded us in a bus powered by steam with a charcoal-fueled boiler on the back end. We rode across town to the district called Pest.

During the ride, I noticed the Star of David and the word "Jude" painted in white on various shop windows. Some of the people that we saw walking around also had a yellow Star of David sewn on the left side of their clothes over their heart. I did not notice any bomb damage from the Allied bombings. After crossing the Danube River, they drove us to the outskirts of town. As the bus turned down one of the streets, we passed a brewery. It was along some railroad tracks. The bus then pulled

Pestvidéki Prison, our so-called "rest camp," where the German sent us to continue our recovery. *Courtesy: John L. Lenburg Collection*

up to a large metal door entrance into a large gray walled area that looked like a prison. This reminded me of the prison that we had back home in Michigan City, Indiana. This was to be our so-called "rest camp" for us to continue our recovery, the Pestvidéki Prison. This was the same prison where the Communists held Cardinal József Mindszenty, a Hungarian Catholic Cardinal, after the war.

The first thing I noticed after entering were people walking around in a courtyard inside the walled area wearing light tan gunny sack clothes with a yellow Star of David on the left side of their breast. At that point, I had a good idea where we were. My worst fear was that they had taken us to a concentration camp. I had heard reports of the Germans sending the Jewish population to concentration camps. This was part of Hitler's plan, "The Final Solution," to exterminate the Jewish population.

They unloaded and took us into a building and told to take our clothes off because they were going to give us a shower. I had also heard news accounts about some of these so-called showers that the Germans gave. I think we all thought that this was it. My heart started pounding. As it turned out, it was a shower. They then moved us to a large building that appeared to be the prison. We went up a metal ladder to the second floor where they put Mike Brown, Alan Barrowcliff, another American airman from another crew and me into a cell. The cell was about eight feet wide by twelve feet long with a large metal door with a peephole and on the outside wall, a small transom-type window. It was possible to see the prison clock from our window. The cell contained a single metal bed frame with a mattress, a table, and a large pail for a toilet.

There was only one metal cot for the four of us to sleep on. We drew straws to see who would sleep on the bed. One of us slept on the

Hungarian Catholic Cardinal József (second from right) who the Communist held at the same prison after the war.

cot's metal springs and another on a straw mattress on the floor. The other two slept on the wooden floor. They allowed us to go outside once a day to walk around in a circle for exercise.

The food consisted of a piece of bread and a thin cup of soup, three times a day. They served our meals in small aluminum pots, which held a little over a half-pint of watery soup. Sometimes, it was hard to tell what the soup was made of, but it was possible to recognize peas (due to the color), cabbage and kohlrabi soups. Each morning, they gave us a chunk of dark bread about the size of half a grapefruit. Prisoners would scratch their names, dates and messages on the aluminum bowls. By doing this, it was possible to tell how long some had been in the prison. The longest time that anyone had marked on a bowl was 108 days and he wrote that he was going "crazy." Bedtime was as soon as the sun went down since there were no electric lights in the cell. There were times when you heard guys screaming. They imprisoned us with criminals, political prisoners, and Jewish prisoners, not just for POWs.

Our cell was infested with bed bugs. At night, they would crawl out the cracks in the wooden floor and bite us. Since I slept on the floor, I would pull my socks up over my pants, turn my shirt collar up and put my hands under my folded arms. This helped keep them from biting us but they would still crawl up on my face while I was asleep. One night, one got into the wound in the corner of my eye and it woke me up. I tried brushing it out, knocking the scab off, and the wound started bleeding. After that incident, I went to sleep by propping myself up in a sitting position against the wall. We put the washbasin on the floor at night, half filled with water, the bugs would crawl into the water in the darkness; and then we would kill them in the morning. I do not know how they ever climbed up the sides of the basin.

We started keeping track of the days by making scratch marks on the wall. I think that we all wondered how long they were going to keep us in this godforsaken place. Fifty years later, I wondered if the Jerries were not purposely trying to play with our minds by doing this. I think maybe that the Germans were trying to soften us up for their interrogation of us.

Exterior, present day, of Pestvidéki Prison in Budapest.
Courtesy: Nándor Mohos

On July 14, 1944, the 15th Air Force bombed Budapest. At about 11 A.M., the air-raid sirens went off and the anti-aircraft batteries started firing. After a few minutes, we heard the slow drone of hundreds of aircrafts, followed by the rain of bombs and loud whooshing sound as they fall from the sky. I think that we all felt the whole prison was going up, with us in it.

All four of us crowded next to the thick iron door away from the window, hoping that no stray bombs would come our way. Even the door vibrated at times from the explosions. It sounded like it was hailing outside, from the falling fragments of the exploding anti-aircraft shells hitting the prison's metal roof. Then there was a violent explosion and everything shook. I thought the prison had been hit; instead, the brewery across the street had been bombed. I think we could almost have gotten drunk from the smell. The beer smell lingered for several days. According to the U.S. news reports that I found after the war for that day, the 15th Air Force bombed oil refineries and the railroad yards in Budapest.

My facial wounds and burns were somewhat slow in healing. I would have to attribute this to poor nutrition. There was a short thin woman in her fifties who came in and checked our injuries. I'm not sure whether she was a nurse or doctor; she wore a long white gown.

One day she came in and looked at Mike. When she pulled a pair of scissors out of her coat pocket, something made a noise and fluttered up in the air between her and Mike. It was a small bird fastened on a string that was tied a buttonhole on her coat. She was very rough while checking Mike's wounds and picked the scabs off his wounds. After that, we tried to conceal our wounds from her.

During my second week in prison, I started getting a boil on the right side of the back of my neck. Our nurse/doctor checked it on one of her visits but did not treat it with anything. This boil was to become quite a problem for me later.

This was a poem scratched on of one of the prison cell's wall:

A Prison Prayer

Lord, guard and guide the men who fly
Through the great spaces of the sky.
Protect them as they take to the air
In morning light and sunshine fair.
Eternal Father strive to save,
Give them courage and make them brave.
Protect them where so ever they go
From shell and flak, fire and foe.
Most loved member of the crew
Ride with them up in the blue.
Drop their bombs upon the foe
But shelter those whom thou don't know.
Guide them well upon their way
Grant their work success today.
Protect them from Hate and Sin
Bring them safely down again.
Lord guide and guard the men who fly
Through their lonely ways, across the sky.

INTERROGATION

On July 22, 1944, the Germans started taking us one, at a time, for interrogation. As I recall, they took Barrowcliff and Brown first, then me. They did not return to the cell after the Germans were finished with them. A Hungarian soldier came to the cell and took me down to the room where they interrogated me. We walked into this room and stopped. There was a German soldier standing by a table and a German officer seated behind the same table. (I thought that he was an officer but, in Barrowcliff's version, he says that he was a sergeant.)

The German officer turned and said something in German to the German soldier standing near the table. He in turn spoke in Hungarian to the soldier that brought me in. The Hungarian soldier brought me over to a chair in front of the table. The officer told me to sit down and offered me a cigarette, which I refused since I did not smoke. A

Captured servicemen leave the interrogation center of Dulag Luft and be transported to various permanent POW camps. *Courtesy: Claudio Michael Becker Collection.*

Hungarian soldier stood behind me. The German officer turned and said something in German to the soldier who had been standing next to the table. After this, the German soldier left the room.

The officer's first question to me in perfect English was what my name, rank and serial number were. I told him, "John L. Lenburg, T/Sgt., 15383056." His second question was what outfit I was with. I answered that under the "Geneva Conventions" I was required to give him only my name rank and serial number. He went on to ask me what the serial number was of our plane and what the purpose was of our June 30th mission. I repeated the same answer to him. He asked me if I would like to spend some more time here in the prison, to which I replied, "No." He asked me again what my outfit was. I gave him the same reply.

Now he appeared agitated and started to get angry since I was not answering his questions with the answers that he wanted. I had the presence of mind to answer him as instructed in my training back in the States. He went through all of his questions again, but I still gave him the same answers. Finally, his face got red and he hit the table with his fist and said that he was going to put me back in my cell. He turned to the Hungarian soldier that brought me in and shouted to him in German to take me back to the "gefangenen zelle" (prison cell).

Then he looked at me. I guess he was waiting for me to start talking, but I did not say anything. The Hungarian soldier did not make a move. At that moment, the other German soldier returned to the room and he told him in German to take me to the "traensport raum."

I knew that this soldier was not going to put me back in the prison cell because when we entered the room he had to speak to the other German soldier, who in turn spoke Hungarian to the Hungarian soldier. The Hungarian soldier did not understand German.

Before I left, the German officer told me that he knew everything about me anyway. He told me the mission that we were on, the plane's serial number, that our call letter was "V" for Victor, my outfit, and more.

After he was finished, I told him that he spoke very good English. He told me he had lived in Detroit before the war and he would be back in the States before me, after the war.

The German officer asked me why I was not fighting on their side since I was of German descent. I did not say anything.

At that, he smiled and motioned for the Hungarian soldier to take me out of the room.

RIDING 40 & 8 STYLE

The Germans hauled me into a room that had other Americans in it. After a long wait, about ten of us were loaded into the back of a military truck. We had four German guards carrying machine pistols. One of the guards was apparently in charge. He ordered the canvas flap covering the back of the truck closed. This way, we had no idea where we were going or if they were going to take us somewhere and shoot us. We did end up being loaded onto a boxcar that sat in a railroad yard. They put us on one end of the car while the guards stayed on the other end. Then told us not to cross the middle of the car or go to the open doors of the boxcar or we would be shot. American officers went to one prison camp, while enlisted men such as me went to another.

Sitting for some time, our boxcar finally moved and attached to a train. Early that afternoon, our train started moving out.

Top: Actual boxcar like the one used to transport POWs. *Bottom:* Exhibit illustrating the transport of POWs by the Germans. *Courtesy: U.S. National Archives and Records Administration*

As you can see, I did not travel in luxury and comfort to the prison camp. The train made periodic stops. They allowed us to get off and walk around under the supervision of the guards. If we had to go the bathroom, we told a guard and then a guard would take us off the train one at a time. We would walk to the rear of the train and then do our business. At some of the stops, the German women would bring buckets of soup to give to the guards and us to eat.

I had no idea where we were. By now, the boil on the back of my neck was getting larger and it was making my life miserable. We arrived in the Vienna marshalling yards in the early evening. It appeared that there was quite a bit of damage to the marshalling yards from Allied air raids. There was a lot of activity going on and they had repair crews repairing the damaged rail lines. The work crews seemed to be slave laborers that the Germans brought in from other countries. There were many bombed-out railcars lying around and the tracks were torn-up. We were very nervous about having to spend any time marshalling yards there. In fact, I am sure everyone was anxious to get out of there. After putting our boxcar on a siding, they reattached it to another train. We all breathed a sigh of relief after leaving Vienna. Incidentally, the marshalling yard was bombed again the next day and again on the July 26 by 500 Liberators and Flying Fortresses. We spent most of the night traveling through Czechoslovakia. The next morning, we went by the oil refinery in Blechhammer Germany. After our unsuccessful mission of June 30, 1944, on July 7-15, the 15th Air Force finally hit the refinery. There seemed to be some bomb damage, but most of the refinery was left untouched.

Early Sunday evening, July 23, our train pulled into the rail yards of Breslau, Germany. The guards took all of us out of the boxcar and walked us into the train station. They brought us to a soup kitchen used to feed German troops that were passing through the station. There was a lot of excitement and activity going on in the station. By now, we were starting to get out of the range of the Allied bombers. I did not see any visible bomb damage. One of the guards took me to a German first-aid station, where they attended to the painful boil on the back of my neck. They put a black salve (a drawing salve) on it and wrapped it with a paper bandage. By now, it had grown quite large—about the size of my fist. My head was starting to pound and I could not lie down and sleep at night. The only somewhat comfortable position to try to sleep was to prop myself up in a sitting position by leaning against the side of the boxcar.

While walking across the station platform, we passed a German youth group coming back from what must have been a weekend outing. The boys were all dressed in their khaki uniforms with short pants, knee high stockings and their red-and-white armbands with a swastika on it. They were marching in step while singing, "Deutschland Liber Alles." It seemed strange in a way to see them. In a way, it felt like I was not here but watching a newsreel in a movie theater back home.

Cathedral Island in Breslau, Germany, after Germans surrendered the city to Soviet forces after a three-month siege in May 1945.

They took us to a different train and loaded us into another boxcar. None of us had any idea where we were going other than to a POW camp somewhere in Germany. The next morning, we pulled into Posen, Poland. The Germans had annexed this area to Germany after they defeated Poland. Here we changed trains again, where they took me to a first-aid station to have a nurse check the boil on the back of my neck.

She put some more black salve on it and rewrapped it. By now, it felt like something was trying to pull my teeth through my head. I still was not getting much relief from the throbbing pain in my head.

While walking across the station under guard to our other train, I noticed three men standing on the platform across the tracks from us. Two of them flashed a "V" sign for victory with their fingers. They did it in such a manner that our guards could not see them. Those of us that observed this were shocked since we were in the heart of Germany. We had not yet heard about the assassination attempt on Hitler's life. I have often wondered who those men were. With our transportation improved, we boarded a passenger train coach to our final destination.

POWs march into the Vorlager of the main camp of Stalag Luft IV. *Courtesy: U.S. National Archives and Records Administration*

STALAG LUFT IV

When we arrived at the rail station of Gross Tychów, we were taken off the train. We were all glad to be off, even the guards. I think that all of us sweated out each marshalling yard for attacks from Allied bombers and fighter planes, while out in the open country.

Late Tuesday afternoon, July 25, 1944, the guards walked us down a small dirt road about two miles to the POW camp, Stalag Luft IV. I found out later how lucky I was that we walked instead of having to run the two miles. This was to be the place that I was held until the war ended. Stalag Luft IV was situated about two and a half miles west of Gross Tychów, in the Pomerania sector of Germany, and south of the Baltic Sea. This was about 54 degrees north latitude, 16 degrees east longitude. If you drew a straight-line west, it would be about even with the southern end of Hudson Bay in the Nunavut Territory of Canada.

The camp was set in a forest clearing about one-and-a-half miles square. The Germans chose this particular forest because of its dense foliage and thick underbrush, which served as an added barrier against escape. There were two, ten-foot-high barbed wire fences surrounding the camp. Rumor had it that the outer fence was electrically charged. Between the two fences was another fence of rolled barbed wire four-feet high. An area 200-foot deep from the fence to the edge of the forest left clear. This made it necessary for anyone trying to escape to cross this area in full view of the guards. Twenty feet inside the wire fences was a warning rail that ran parallel with the barbed wire fence. If a prisoner touched or stepped across this rail, the Germans would shoot them.

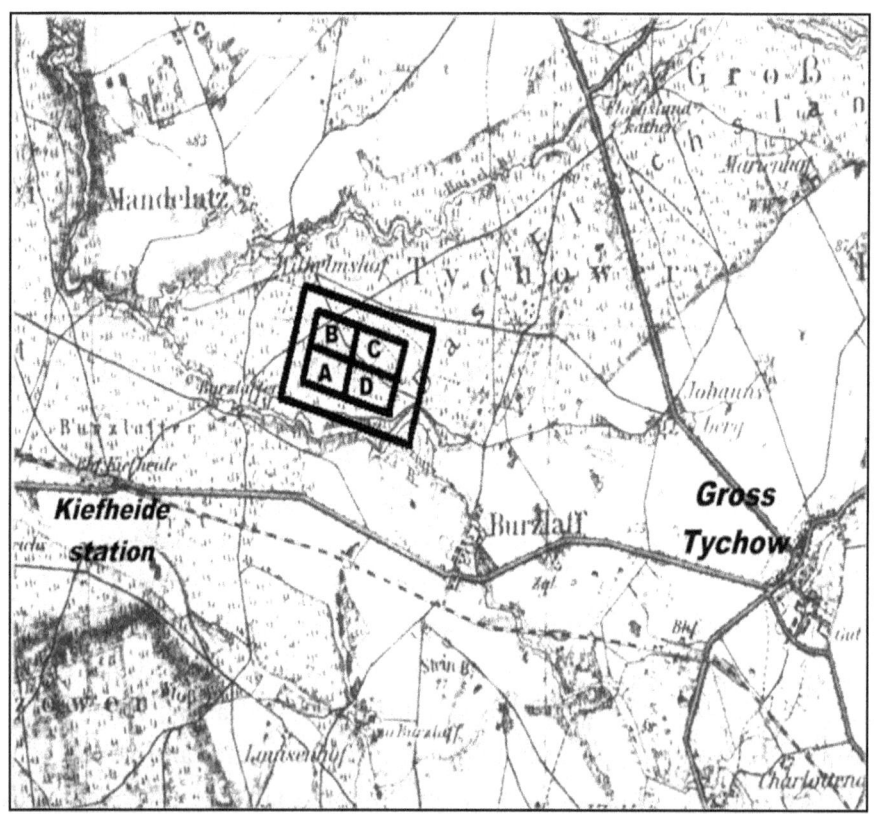

Map location of Stalag IV about two and a half miles west of Gross Tychów.

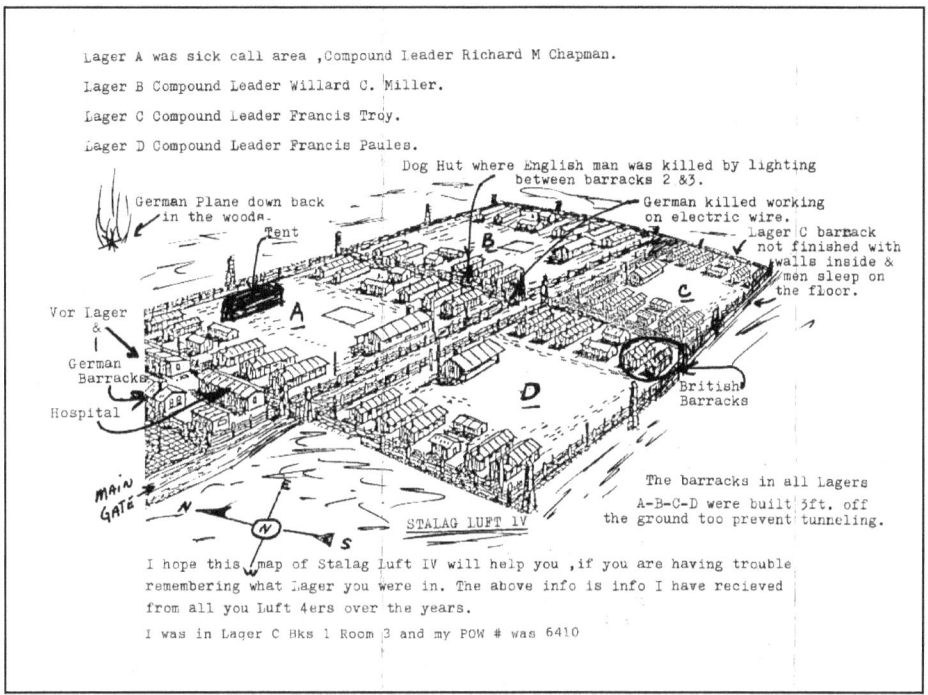

Drawing of the general layout and compound of Stalag Luft IV. *Courtesy: John L. Lenburg Collection*

Watchtowers, built at close intervals around the camp, were equipped with powerful spotlights and machine guns. The barracks, built off the ground, were in the open, so the guards would have a clear view.

This camp, known as "Strafe Lager" or punishment camp, was run by German officers who hated American or British airmen. No one had successfully escaped from the camp, although prisoners had made several attempts. Those who attempted were shot, killed, and buried at the edge of the clearing. Their graves were marked with white crosses in full view of our compound.

The Germans had learned a lesson on how to build an escape proof from the escape problems that they had at Stalag Luft III.

This prison camp opened in April 1944 with the transfer of 2,500 POWs from Stalag Luft VI, known as Heydekrug in East Prussia. Most of these prisoners were Americans, except for about 500 non-

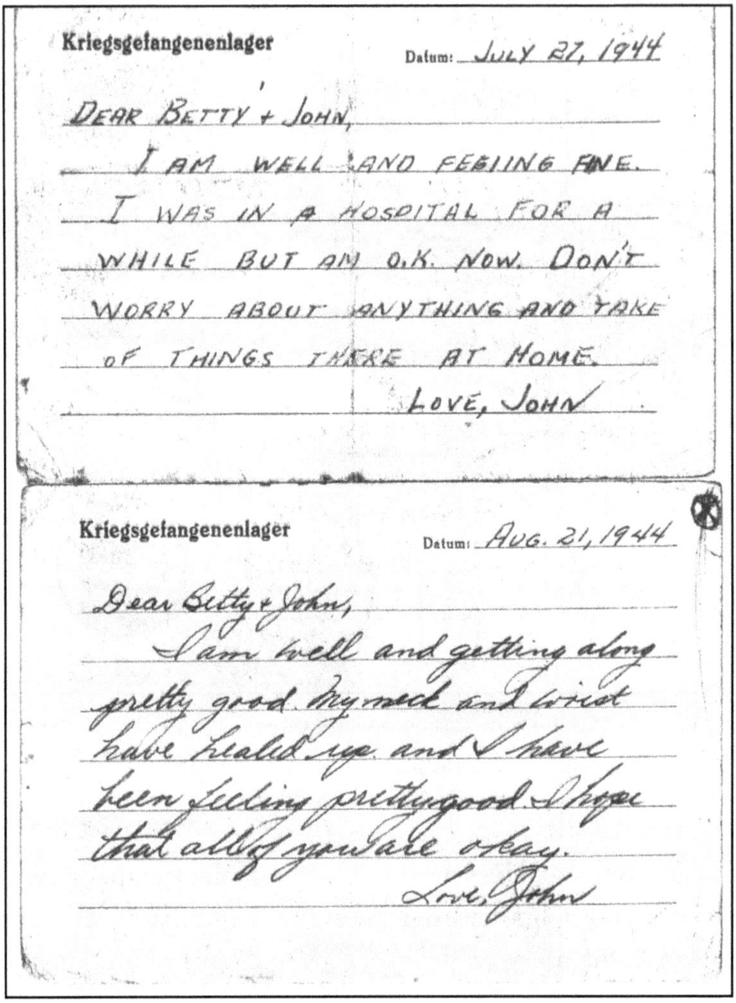

> **Kriegsgefangenenlager**   Datum: JULY 27, 1944
>
> DEAR BETTY + JOHN,
> I AM WELL AND FEELING FINE. I WAS IN A HOSPITAL FOR A WHILE BUT AM O.K. NOW. DON'T WORRY ABOUT ANYTHING AND TAKE OF THINGS THERE AT HOME.
> LOVE, JOHN

> **Kriegsgefangenenlager**   Datum: AUG. 21, 1944
>
> Dear Betty + John,
> I am well and getting along pretty good. My neck and wrist have healed up and I have been feeling pretty good. I hope that all of you are okay.
> Love, John

Letters I wrote in July and August 1944, to my aunt Betty and uncle John Lenburg. *Courtesy: John L. Lenburg Collection*

commissioned British RAF officers. At the time that I arrived here, only lagers "A" and "B" were in use. Lager "C" and "D" were still under construction but just about finished.

On July 14, 1944, 2,500 prisoners from Stalag Luft VI prison camp in Heydekrug, East Prussia were jammed into the holds of the cargo vessels, Insteberg and Masuren, at the seaport of Memel and transported down the Baltic Sea to the port city of Swinemünde. Conditions in the holds were so crowded that only half could sit while the other half had to stand. They were without food and water for three days. After docking three days later, on July 17, 1944, they were immediately loaded into waiting boxcars, shackled hand and foot with leg irons and chains in pairs, and promptly shipped to Keifheide in Germany.

The next day, on July 18, 1944, they were unloaded from the boxcars and made to stand in the blazing July sun. As they moved out, a redheaded German officer Hauptman Pickhardt, who was the captain of the guards, gave the command, "Quick March!" Young German Kreigesmarines with fixed bayonets forced them to run the two miles to the camp. During the "Run," as it was called, Pickhardt incited the guards by yelling, "American airmen were gangsters who received a bonus for bombing German children and women." As soon as a Kriegie would fall, a guard would rush in with his guard dog, letting the dog bite the Kriegie on the arms and legs. German guards clubbed other POWs with their rifle butts in the groin or shoulders or bayoneted them. Some of the prisoners were in such a terrible physical state that their fellow POW carried them along in shackles.

The Germans had set up camouflaged machine gun posts along the road hoping that the prisoners would bolt for the woods when this action took place. But not one man left the formation. In fact, there were many acts of heroism that day. Those responsible for this atrocity, the German officers and guards from Stalag Luft VI, were ultimately charged for war crimes after the war ended. Resulting from the dog bites and bayonet stabbing, one man had to have both of his legs amputated.

Welcome to Stalag Luft IV. I arrived seven days after this event took place. Some groups of POWs that arrived after us, the Germans also made run to the camp. Since we walked, we were very, very lucky. I did see the camouflaged machine gun posts along the edge of the road.

Example of a typical POW barrack: At Stalag Luft IV, the barracks were so crowded that there was not ample room for all the POWs to bunk together, so they took turns using the floor on which to sleep. *Courtesy: U.S. Air Force Academy Library*

KRIEGIE #6410

The Germans took us to a building outside of the barbed wire enclosure that was called the Vorlager. There they told us the dos and don'ts of the camp and had our mugshots taken. I became Kreigsgefangenen, number 6410. Most of the regulations were don'ts or the Germans would shoot you.

I met two of the German guards who had a part in the "Run." They were Oberfeldwebel Fahnert ("Iron Cross") and Feldwebel Schmidt ("Big Stoop"). They would physically abuse prisoners, as I saw in one case. Big Stoop picked up one of the POWs and tossed him over a partitioned wall. He also liked to cuff a prisoner on the ear with his open hand using a sideways movement. This would cause pressure on the eardrums, which would cause it to puncture. I tried to avoid them as much as possible.

Fahnert was nicknamed "Iron Cross" because he always wore the

German Iron Cross medal (awarded to all ranks usually for valor) around his neck. We affectionately called Schmidt the "Big Stoop" because he was about six-foot, seven inches tall, had very large hands, and a mean look about him. The Wartime Crimes Commission would want both of these men after the war was over. In November of 1945, they arrested Fahnert for ill-treatment of POWs, for looting of Red Cross parcels, and for the treacherous and unprovoked shooting and killing of POWs, Sergeants Walker and Niles.

The Germans put me in Lager "A" temporarily, since the other lagers were not completed. They housed some of the British POWs in tents since the barracks were full and put me in with them. From these POWs, I learned about the "Run." They put me next to a RAF sergeant who had been a pilot and was shot down during the invasion of Holland in 1940. The Germans issued me eating utensils and a bowl, two thin Luftwaffe issue blankets made of horsehair, and used G.I. clothes to wear.

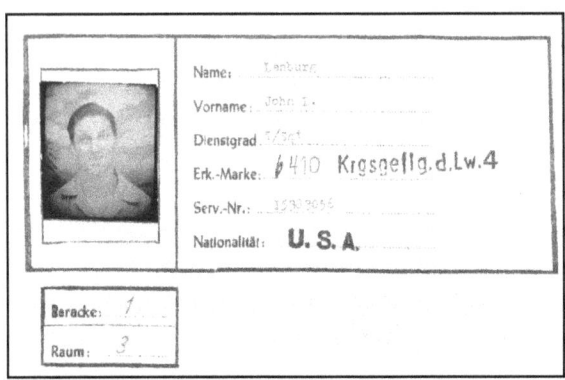

My POW I.D. at Stalag Luft IV. *Courtesy: John L. Lenburg Collection*

Jack Nagle was living in one of the barracks in Lager "A." He had arrived several weeks earlier since he had not been in the hospital. It was good to see him again. Jack told me the Germans had kept him in solitary confinement while he was in the Budapest prison. His only food was bread and water twice a day.

The first thing that I did was to go on sick call to see if I could get this boil on the back of my neck taken care of. By now, it was almost the size of a small grapefruit and I could not turn my head anymore. The doctor that checked me was a POW from New Zealand. Removing part of the core, he put some more black salve on it and rewrapped it with another paper bandage. He told me to come back the next day. I did before having the rest of the core removed. The core was as long as my little finger and as round as a dime. He told me that it was a good thing that the Germans had treated it.

If they had not, I might not have made it. During the time that I was on sick call, I ran into some of the casualties of the "Run." One of the British POWs had fifty-seven bayonet stab wounds; another had to have both of his legs amputated. Others underwent treatment for dog bites besides bayonet wounds.

On Thursday, July 27, 1944, the Germans made all of the POWs line up. We stood outside in the middle of the compound for hours. The Gestapo checked us one by one with the card index and file photo they had on each one of us. It was then that I learned about the attempt on Hitler's life. After this, things changed at the camp. German Luftwaffe personnel up to this point had run the camp. The Wehrmacht brought along with them the Gestapo. I guess no one trusted anyone. Then we got a new German interpreter. Kriegies nicknamed him the "Green Hornet" since he was in the German Wehrmacht and was dressed in army green.

My barracks at "A" Lager and the back of barracks 1, 2, and 3 at Stalag Luft IV.
Courtesy: John L. Lenburg Collection

My Bible and hymnal and prayer book from my time as a POW. *Courtesy: John L. Lenburg Collection*

LIFE IN KRIEGIELAND

On September 27, 1944, other American POWs and I were moved to Lager "C" and the British to Lager "D." The Germans put me in Barracks 1, Room 3. Every barracks had twelve rooms and a latrine. Each room was about fourteen-by-sixteen foot and had eight triple-decked bunk beds, with shredded wood mattresses supported by four bed slats. There were twenty-four men in each room. We had a little table and a small pot-bellied stove. During the winter, each room was given twelve peat bricks a day for fuel. You had to dress warm.

The following are the names of American POW airmen that were in Barracks 1, Room 3:

T/Sgt. John H. Anderson	37502787	4005
Sgt. James R. Brown, Jr.	33627687	4025
S/Sgt. William N. Brower	42016172	3965
S/Sgt. Jack R. Burton	14003542	1580

Red Cross volunteers pack parcels destined for POWs in enemy prison camps in 1943. *Courtesy: Office of War Information Photograph Collection (Library of Congress)*

S/Sgt. Ernest Crawford	34723090	4209
S/Sgt. Francis W. Foltz	36123565	1589
T/Sgt. Richard R. Hair	36432382	6534
Sgt. Donald Hite	36596791	7291
Sgt. Harold M. Hoover	15113441	7189
T/Sgt. John L. Jones	14119473	7203
S/Sgt. Ernest C. King	17019789	7211
S/Sgt. Stanley Kudlo	36590847	7212
T/Sgt. Edward J. Lafferty	39325960	7213
T/Sgt. John L. Lenburg	15383056	6410
S/Sgt. Drexel D. Lange	37476570	7214
Sgt. Ira E. Lewis	39298268	7217
T/Sgt. August W. Maurer	36804761	4237
S/Sgt. Charles R. Merritt	14170027	6587
T/Sgt. James I. Meyers	15383261	4123
S/Sgt. Donald L. Nelson	37552876	4132
S/Sgt. G.P. Lynch	11058136	5020
S/Sgt. N.A. Lynch	15108258	5027
S/Sgt. Gerald F. Foretich	14033355	2435

Red Cross packages to supplement POWs diets await shipment. *Courtesy: The International Red Cross*

After moving to the new lager, Kreigie life settled down to the daily routines. Hauptman Weinert was the German officer in charge of our lager. He was a repatriated pilot who had shattered his leg in an airplane crash. Weinert walked with a slight limp, spoke English, wore a long blue leather coat and black boots and demanded a salute from the POWs and guards. He did not use the stiff-armed Nazi salute, instead what we called the soft Nazi salute of just raising up his right arm. He always carried what looked like a riding crop clutched tightly in his left hand that he held behind his back. He stood very erect and had all of the appearances of a Prussian officer.

We had two meals a day from our kitchen, two roll calls a day and that was about it. Occasionally the Germans would have a lock out, when they would go through the barracks looking for contraband. For recreation we played baseball, football, walking, read books, played cards, checkers, chess, and more.

Occasionally we would get Red Cross food parcels to help supplement our diet. Red Cross parcels had the following items:

Powdered Milk	16 oz.
Processed cheese	8 oz.
Margarine	16 oz.
Corned beef	12 oz.
Spam, canned	12 oz.
Liver paste	6 oz.
Tuna or salmon	8 oz.
Prunes or raisins	15 oz.
Biscuits, K-ration	7 oz.
Chocolate bar	8 oz.
Coffee, Instant	2 oz.
Salt and pepper	1 oz.
Jam	6 oz.
Multivitamin tablets	16 tablets
Sugar, lump	8 oz.

LIFE IN KRIEGIELAND

Sample Red Cross parcel sent to American POWs in camp.

We were supposed to get one of these parcels per man per week but we were lucky if we got a half of one once a month. I always had to share it with one or two other persons. Consequently, we ate very sparingly, as the Germans did not give out parcels on a regular basis. I guess the guards, camp commandant and a few other people were stealing the parcels.

For breakfast, we had hot water or ersatz coffee (made from barley). Anything else that we ate would have to come from our Red Cross parcels. Lunch or dinner could be soybean soup, sauerkraut, boiled potatoes or some other kind of soup. Occasionally you might find a little horsemeat in the soup. This was about our only source of protein outside of our Red Cross parcels. We also received a ration of eight loaves of German black bread every week per room. This broke down to one-third of a loaf per man. It was of a rather heavy texture, dark brown in color, and made from a mixture of wheat hulls and wood pulp that had been chemically broken down. When cutting a piece of the bread to eat, you would slice it very thin.

The Germans allowed us to walk around our compound or lager, as long as we did not cross the warning rail. The guards in the towers made sure of that. At night, they locked us in our barracks at sundown. The guards shut the front and back door, putting a bar across the outside to prevent it from opening from the inside. Each room had a double window that would swing out to open. At night, they shuttered and barred the windows from the outside world. With our barracks built up off the ground, it was impossible to dig a tunnel without the German guards who were on duty seeing. At night, guards with dogs walked the compound, and the tower guards probed the grounds with their spotlights. A small transom above each of the doors left open at night, let air in. Of course, the Jerries tried to build this prison escape proof after the problems they had with escapes from Stalag Lufts III and VI. Stalag Luft III was the camp from which the great escape was made.

The one utensil given to me by the Nazis--a spoon with the Nazi insignia engraved on the handle.
Courtesy: John L. Lenburg Collection

Life became boring at night. All everyone talked about was food and his plans on what he was going to do when he got out of the service. We had the saying, "You could plan all day and dream all night long, but in the morning, you still woke up in the same place."

The days, weeks and months seemed to drag by. Some of the POWs could not handle this confinement and went "around the bend or over the hill" as we called it. One in particular tried to climb the barbed wire fence that enclosed the camp and stretched for miles. The Germans shot him. Others they dragged out of our lager screaming.

Diversions were few, except for several incidents that happened. On September 20, 1944 when a Fw-190 crashed in a wooded area near the camp. Three Fw-190s would practice dog fighting over our camp. This day, the sky had a low overcast and we could hear them practicing dog fighting above this. Suddenly one dove down through the overcast and leveled off at treetop level. The second one followed, right behind the first. Then the third one descended, but he did not level off in time. He dove right into the ground, causing a big explosion.

All of us Kriegies stood around and cheered. This upset the guards in the tower and they started firing their machine guns into the compound. We all hit the deck after which they warned us to knock off the cheering.

Another diversion was watching the V-1s and V-2s that the Jerries launched from their rocket base at Peenemunde, which was about thirty miles away. We could see the vapor trails left behind as the rockets climbed up in the sky. The RAF made a couple of night raids on this facility. On the night of November 15, 1944, there was an air raid on this facility. The air-raid sirens sounded and the camp lights went off. The RAF lit up the sky with their flares and bombs. Even though we were locked in the barracks thirty miles away, we could observe the sights by looking through the open transom window above the door.

Speaking of rockets, the Germans had a large tank that they used to bring in our compound to excrete the human waste material from the latrines. We called this device "The Honey Wagon" or "Hitler's V-1" and it was manned by Russian POWs. The tank sat on a metal frame with four rubber-tired wheels. A tractor that ran on a charcoal-fed boiler pulled it. The Germans would use this human waste as fertilizer and spread it over their farmlands.

One day, they brought this V-1 into compound, which the Russian POWs readied to suck out the human waste material from our latrine sump. In getting this device ready, they would pump an explosive gas

On September 20, 1944, a German Fw-190 crashed in the woods near camp.

into the tank, attach a large hose to the rear of it, while putting the other end down into the cache basin. Next, they would turn some valves on and light a small fuse in the rear of the tank. Upon doing, there would be a loud "Whooom!" as a small round door on top of the tank slammed shut. When lighted, the gas ignited causing a vacuum, which drew the waste into the large tank. Everyone in camp would jump when this explosion occurred.

Then, one day, several of us happened to be watching this process-taking place. One of the Russian POWs smiled at us as he lit the device at the rear of the tank. At that same moment, the cover on top closed; the whole tank caved in with a very loud bang. Apparently, the Russian, on purpose, did not turn one of the valves on and the outside air pressure crunched the tank. Of course, there was all kind of commotion with the German guards running around hollering and screaming. The Russian POW just shrugged his shoulders at the guard, as if he did not understand what had happened. After that episode, the

Germans used ox-drawn wagon with a wooden tank on it. With this device, the Russian POWs had to hand pump out the waste material. It was similar to an old-fashioned fire wagon with a long pump handle on each side of the tank. We called this the "Little V-2," another one of Hitler's secret weapons.

Occasionally the guards would test fire their machine guns in the guard towers. When this was done, they would empty the barracks of all the Kriegies. They would herd us all over to the side of the compound under the guns they were testing. Then they would start firing their guns. Several times, we found bullet holes in the barracks.

The Jerries conducted search after search looking for our secret radio that was hidden somewhere in our camp. Some of the POWs had smuggled it into camp from Stalag VI. The BBC would broadcast news reports at certain times of the day meant exclusively for POWs. A runner would carry the news from barracks to barracks. While the

A drawing of "The Honey Wagon" or "Hitler's little V-1." *Courtesy: John L. Lenburg Collection*

runner was giving us the news, we posted guards to be on the lookout for any Jerry guards. I am sure that this drove the Jerries crazy looking for the radio that they never found. Once a week they gave us a German newspaper called, *Allgemeine Zeitung.*

It was already fall but there were no leaves turning color because our camp was in the middle of a pine forest. The weather was getting colder since we were in the northern part of Germany, near the Baltic Sea. Soon it would be Christmas and, of course, winter, and winter meant snow. Some of the Kriegies really had a hard time adjusting to confinement in the POW camp, especially with the holidays coming up. Our escape committee advised us not to try to escape but just sweat the war out now. Some of us just kept telling ourselves repeatedly that this could not last forever. I really do not know how some of those British POWs could handle being a prisoner for four or five years.

At the edge of the clearing next to the pine forest were four white crosses. These were the graves of the four POWs, two of them shot dead by the guards. They were buried so we would be able see the graves when we walked around the compound. Lt. Col. Aribert Bombach was the commandant of the camp. Captain Pickhardt was one his assistants, who was in charge of the guards and security. Along with some of their henchmen, they had mutual contempt for American airmen and the Geneva Conventions. Bombach spoke English, was a Nazi Party man, and served as a pre-war agent in France. Pickhardt was a fanatical Nazi, known as the "Mad Captain" or "Butcher of Berlin, in charge of the "Run" who as previously mentioned physically abused POWs. The War Crimes Commission later charged them, along with some of the guards, for crimes against POWs.

Our camp leader, Richard Chapman, protested the continued abuse of newly arriving prisoners. The Swiss legation representative, who visited our camp for the International Red Cross, filed complaints with Lieutenant Colonel Bombach for these violations of

Article 2 of the Geneva Conventions, rules that apply to prisoners of war in times of armed conflict. Article 2 states:

> *Prisoners are in the power of hostile governments, but not of the individuals. They shall at all times be treated humanely, particularly against acts of violence...and measures of reprisal....*

Bombach refused to acknowledge the POW complaints, so consequently these protests were in vain.

The cover of *Yank* magazine from its December 24, 1944 edition spreading the spirit of Christmas to American servicemen. *Courtesy: National World War II Museum*

CHRISTMAS

A Kriegie in Lager A tried to escape by running for the fence. Someone pulled him back, so the guards never got off a single shot. He went off his rocker. The term we used when this happened was he went "round the bend" or "stir crazy."

Four days before Christmas, at about 1 A.M., the air sirens went off making a loud wailing noise. The RAF was over Stettin again. The explosions kept us up most of the night, even though it was thirty miles away. Everyone watched what activity that was visible through a window above the doorway. The Germans warned us afterward that if anyone was caught looking out the window during an air raid, if we did this again, they would be shot. Christmas, 1944 was certainly different from any that I had ever experienced before.

According to our various news reports, the ground war was going well and the air war over Europe was really in full swing. From what we gathered, maybe liberation was not far away. We could tell downed airmen were pouring into the prison camp by the droves. They also brought the latest news reports. The number of POWs had climbed from 3,000 when I entered the camp to almost 10,000 by the end of January 1945. The Jerries were working feverishly building Lagers "E" and "F." The word was, "Stay put and hang on, no escape attempts."

The weather had turned bitter cold with temperatures below zero with ice and snow. There were twenty-four guys huddled in our little rooms with our little pot-bellied stove trying to keep warm. As noted before, the Germans issued each of us two blankets made of horsehair. I swear if you held them up to a light that you could see through them. Consequently, we usually slept with our clothes on.

On Christmas Eve, 1944, they permitted us to walk around the compound after dark. We were on parole, so to speak, on the promise that we would not try to escape. We visited with other POWs in other barracks in our compound or other lagers until 1 A.M. Our kitchen personnel had been saving pieces of beef for some time and made everyone in our compound small hamburger patties to celebrate. The hamburgers were great,

Original drawing from the Wartime Log of fellow POW Willard Miller, Stalag Luft IV. *Courtesy: The National World War II Museum*

Back home in the U.S.: The December 1944 *War Prisoners Aid News* offering hope to American prisoners of war.

although we all thought that it was horsemeat. We savored every morsel.

Some of us had been saving our raisins from our Red Cross packages from which we made "Raisin Jack" to pass around the compound on Christmas Eve. Other POWs made Kriegie cakes, using water, powdered milk, margarine, sugar, and crushed graham crackers from our Red Cross parcels. We put mixture in a KLIM can to bake it. If you are wondering what a KLIM means, it was milk spelled backwards. Powdered milk from our Red Cross parcels came in these cans. We used the cans for everything.

Dr. Knud Christiansen from Denmark, Sweden visited our camp. He was with War Prisoners Aid sponsored by the YMCA. He brought many recreation articles, as well as cards and books to us. The Germans handed out Christmas Red Cross parcels. This parcel contained one pipe, tobacco, cigarettes, mixed nuts, candy, fig bars, washrag, honey, butter, tea, two pictures, variety game, roulette, cards, turkey, cheese, Vienna sausage, pudding, bouillon cubes, deviled ham, chewing gum and more. The guys were like little kids opening up their package. Groups of Kriegies walked around the compound singing Christmas carols. The guards went around wishing everyone a Merry Christmas. We had no roll call Monday Christmas day.

Dr. Knud Christiansen of War Prisoners Aid from Denmark, Sweden, with his wife Karen, who visited camp. *Courtesy: The Jewish Foundation for the Righteous*

On Tuesday, December 26, 1944, everything was back to normal. A few days later, we heard the news about the German breakthrough and the Battle of the Bulge was on. The Allies were in retreat. No wonder the guards were walking around with big smiles. I think that was our lowest point. By mid-January 1945, Allied Allies had stopped the German offensive and recovered most all the area lost in the German attack.

- 1945 -
THE LONG WALK

I
n early January 1945, the Russians had started their winter offensive from Warsaw, Poland. The German High Command issued orders to evacuate our camp. Stalag Luft IV was northwest of Warsaw. The Russian breakout initiated the evacuation of Luft IV. By February 3, 1945, the front lines were forty miles south of Luft IV and extended to the Oder River. The only route left for our evacuation was northwest through a narrow fifty-mile gap to Swinemunde, near the Baltic Sea. If the Russians kept advancing as they were, our camp would be right in their path. Little did we realize at that time the hardships that we would be forced to endure before the war ended. The weather here had been very cold with quite a few heavy snowfalls.

On January 15, 1945, three thousand sick and crippled POWs were loaded into boxcars and shipped to various POW camps in Germany.

January 5, 1945: A Soviet T-34 tank advances across the highway Zhitomir-Berdichev, the first Ukrainian front, during the winter offensive in Warsaw, Poland, forcing the Germans to order the evacuation of our camp. *Courtesy: World War II Today*

On February 5, 1945, the Germans put the rest of the camp on notice. They told us that they were evacuating us to a safer camp, for what was a three- or four-day march.

Preparation for the march included making bedrolls, knapsacks, taking KLIM cans and packing warm clothing. It was common to see small groups and individuals walking a fast pace around the compound. This was to try to condition themselves for the rigors of marching. Our inactivity in camp had left us in poor physical shape. Therefore, we had a lot of catching up to do.

We would need a utensil for drinking, for cooking and one for eating. These utensils had to be light, easy to pack, unbreakable, and able to withstand minimal cleaning. This is where the KLIM cans came in handy. The KLIM can was milk spelled backwards and came in our Red Cross food parcels. It contained powdered milk. It was approximately four inches in diameter and about three inches deep.

Some of us took our blankets and made bedrolls by folding them in half and sewing the bottoms and sides up. We left the tops open. Then we took and pushed one of the sewn blankets into the other. By doing this, we made the bed, so we crawl into it. Next, I took extra shoestrings and tied them together. Then I rolled up my blankets with what food and personal belongings I had, which was not much. Using shoestrings, I made a sling by tying an end to each end of the rolled-up bedroll. We also used shirts to make knapsacks to help carry some of our meager belongings by tying the sleeves together, plus sewing up the bottom. Believe me, there was a lot of confusion in the barracks and in our room, with twenty-four guys trying to get things ready made for the march.

On February 6, 1945, at about 10 A.M., the guards entered our compound blowing their whistles and yelling the familiar, "Raus! "Raus!" The Germans ordered us outside for roll call and made to line up with whatever meager belongings we could carry. The Jerries had us turn and start walking out the main gate in a line of four abreast. As we moved out of the camp past the Vorlager, they gave each POW a full Red Cross parcel and a third of a loaf of bread. It was the first time that any of us had received a whole Red Cross parcel, which might have to last us a long time. This marked the beginning of the full evacuation of the remaining seven thousand POWs.

The twenty-five hundred POWs of Lager "C" were the first to move out the gate with our guards carrying rifles, machine pistols and their K-9 corps. Hauptman Weinert was the German officer in charge of our group and Capt. Leslie Caplan, a captured American flight surgeon from the 449th Bomb Group, was in charge of attending to the sick. The Germans told us that we would walk about two weeks and then set up camp at an abandoned sugar factory. Little did we realize this was beginning of what they would later call "The Death March of Stalag Luft IV." Some of us would not be around to see its end.

The next eighty-plus days that we were on the road, we covered more than six hundred miles. During the walk, we froze and became infested with lice, were plagued with dysentery, diphtheria, typhus and pneumonia.

We would live in filth, sleep in barns with farm animals or empty fields and dodge aerial strafing. Then we were marched from the eastern front to the western front and back to the eastern front. For food, we would average seven hundred and seventy calories a day on German rations for the first fifty-three days of the march. If it had not been for the Red Cross food parcels we occasionally received, many more of us would have perished. These parcels gave us an additional six hundred calories a day. They were the only appreciable source of protein for us. The last thirty days of the march, the Germans gave us a little more food and Red Cross parcels. They became a little more humane in their treatment of us.

In the beginning, all you could hear was the clinking of the KLIM cans hanging from the Kriegie's waist. To this chorus everyone seemed to be in step. After a while this changed to just a bunch of clinking cans. Our overnight stays and sufferings—the physical and verbal abuse, the hunger, illness without adequate medical treatment, improper clothing to protect us from the extreme cold weather—all became part of our daily life.

On February 13, 1945, we arrived at the small town of Dobberphul. We were now eight days out of Stalag Luft IV. It was obvious that no end of our marching was in sight and our destination was still unknown. Stalag Luft IV was one hundred and forty-three miles behind us. We had settled in to the daily routine of walking, sleeping, freezing, and starving. We were awakened at 5:00 A.M. every morning with the familiar "Raus! "Raus!"

After roll call, we had breakfast consisting of two or three boiled potatoes, depending on the size, and then marched in columns of threes or fours by following the leader to our next destination.

February 14, 1945 was a gray and dismal day with a constant freezing rain and the relentless freezing cold. After marching for two hours, the Germans gave us one five-minute rest period. We were supposed to complete our toilet duties and take a rest in this short period. In the freezing weather, this was impossible.

The usual procedure at dusk was the separation of the column into small groups and then distribution to whatever barns were available, with the rest bedding down in the open fields. However, on this particular day, there was no separation of columns or distribution to available barns. Darkness settled in and we were still marching. Our only food was the few potatoes we received at 5 o'clock that morning or whatever we had left from our original Red Cross parcel issued eight days before. After sixteen hours of marching, we arrived at our resting place.

During the walk, if you fell behind, the guards prodded you with their rifles to keep up with the main column. If you fell behind consistently, a guard would fall behind with you. About five or ten minutes later, the guard would rejoin the column without the POW. This usually meant that the POW was most likely shot and his body discarded alongside of the road.

The Germans put the very seriously sick on a horse-drawn wagon that followed our marching column. Captain Caplan was he doctor in charge of this wagon, pulled at times by the POWs. The same farm wagon used to haul manure.

The doctor repeatedly pleaded with Weinert to hospitalize the sick and dying but every time they turned him down. If a person died, the body was removed and left behind. Immediately, someone else took his spot, as we had a long line of men following the wagon waiting to get on. In short order, it became known as the "Death Wagon." Once they put you on this wagon, the only way that you got off was feet first.

Some of the POWs seemed to give up hope at this point, so it was better if you could force yourself to keep walking.

Dysentery was our biggest problem since food and water were very poor. For dysentery, we would burn wood, and then eat it like charcoal, which helped. Other problems were pneumonia and open foot sores that would not heal. Then the dreaded disease typhus began to raise its ugly head. We also had some cases of diphtheria and tuberculosis.

The following excerpt is from an article Captain Caplan wrote for *Air Force Association* in August 1982 about the serious health and medical issues we POWs faced:

As a medical experience, the march was nightmarish. Our sanitation approached medieval standards. The inevitable result was disease, suffering, and death. We soon found out what it means to live in filth on low rations and little water.

Our first problem was handling the stragglers. Volunteers at the end of the column would spot a fellow who was weakening and support him. When a straggler could not keep up even with help, a medic would stay behind to give him protection of his Red Cross armband. In that way, the straggler was much less likely to be bullied by the guards. Sometimes it didn't work, and both the medics and stragglers were gun-butted.

The number of stragglers increased daily, and it became impossible to support them all. The Germans gave us a farm wagon to carry the sickest.

At first the stragglers consisted of men with blisters, aching feet or joints, and tired muscles. The medics made a slogan: "Keep on marching and your blisters will turn into calluses and your aches into hard muscles."

All too soon the straggling became more serious. Blisters became infected and many ugly abscesses that developed required treatment to open them. The mud and cold turned to frostbite and, in some cases, gangrene and amputation. The fifth day of the march, the Germans diagnosed the first case of diphtheria, followed by a case of erysipelas. Then cases of pneumonia began showing up and, on the latter part of the trip, tuberculosis.

The illness that really plagued us was dysentery, a natural sequence of living in filth. The dysentery overwhelmed us. Almost everyone had it. Day after dreary day we marched along, the roads lined with our dysenteric comrades relieving themselves. It was a sad spectacle seeing a soldier relieving himself right in a village street, but this was a common sight.

As we neared the area of the port of Stettin area, there was a lot of activity.

The Volkssturm (People's Army or Home Guard) were building traps and setting up machine guns. I guess that they were going to try to stop the advancing Russian Army. Late in the afternoon, the Germans brought up a portable soup kitchen. It was the only time during the entire walk-through Germany they did this.

The Jerries had us bed down in an open field. I was sure glad that I made my bed roll. I had teamed up with one of my roommates. I did this since we had to share our Red Cross parcels with someone else. His name was Donald Hite, a native of Missouri. We shared food parcels back at the camp and kind of looked out for each other. This we called a combine.

That night, the RAF flew over, dropping flares and bombs. A few hundred yards away from us was an anti-aircraft battery. Of course, they were firing these until the raid was over. The noise was deafening.

Captain Caplan, in his testimony at the War Crimes Office, Civil Affairs Division December 31, 1947, best described our conditions:

> *On 14, February 1945, Section C of Stalag Luft IV had marched approximately thirty-five kilometers (21.74 miles). There were many stragglers and sick who could barely keep up. That night the entire column slept in a cleared area in a wood. It had rained a good bit of the day and the ground was soggy; but it froze before morning. We had no shelter whatever and were not allowed to forage for firewood. The ground we were to sleep on was littered with feces from prisoners who had previously stayed there with dysentery. There were many barns in the area but no effort was made to accommodate us there. There were hundreds of sick men in the column this night. I slept with a POW who was suffering from pneumonia. Some of the men collected tree branches and placed them in layers on the ground to protect them from the feces and the soggy ground.*

A German overseer, with conscripted farmhands, usually ran the farms where we stayed. The farmhands were men or women conscripted from the countries Germany had over run. We called them slave laborers.

If you had to go to the bathroom at night, one of the first rules were to remove your shoes, then crawl to the barn door over many sleeping Kriegies. Next thing was to attract the guard's attention. When you got his attention, you would say, "Posten! Posten! Ich haben zu shitzen." Only after the guard gave you permission could you leave the barn. It got to be quite a problem being locked in a barn at night with five-hundred Kriegies all having the G.I.'s.

One of our first setbacks was the arrival of some lousy little parasites called lice. Webster's describes them as a "small, flat, wingless parasitic insect, with either biting or sucking mouths that infest the hair or skin of men or other warm-blooded animals." As POWs, the lice selected us as the warm-blooded animals since we had bedded down with the farm animals.

On sunny days, you would see POWs strip down to their waist to remove the eggs usually found in the seams or the weave of their clothing. They never seemed to be content to feed in one place. Their travels carried them from our necks down to our waists. We would try to kill them by pinching or squeezing them. We always seemed to lose the battle and

Capt. Leslie Caplan (far right) with (from left to right) Francis Troy, Hauptman Shetter, and John Kohl. *Courtesy: Greg Hatton, b24.net*

conceded that if you killed one, a thousand would come to his funeral. It became a challenge just to meet the next day.

A fellow Kriegie Joe O'Donnell wrote the following in his diary on the 35th day the march:

> *I sat in silent solitude and stared at my reflection in a dirty, ice-encrusted pool of water, filled by the melting snow. In it, I saw a harried, starved, unshaven and unbathed skeleton of a person that was myself who had once walked with pride and dignity as my companions did. Now I walked with animals, like myself, as companions. I now urinate and defecate in the woods, like an animal, with nothing more to wipe with but a leaf, some straw or my hand. I now urinate in the streets of small towns like a dog. I finally realized that the thin, pale, wild-eyed creature with death like expressions that I was seeing in the reflection of the pool of water was I. I no longer needed the pool to reflect my degradation. Day 35 was my day of mental depression and; Little did I realize that this mental depression and anguish would continue for another 51 days.*

A POW John O'Connor from the 460th wrote this account:

> *My experiences were the same as thousands of other Prisoners of War on the Death March: Starvation, dysentery, frostbite, exposure to the bitter cold and exhaustion. The miracle of the next steps you never thought you could physically take. The mental torture, wondering if you could meet the challenge of the next day and sometimes wishing you would not wake up to face it. Squatting at a slit trench feeling the blood pouring out of your bottom. Sleeping in barns, where a group of prisoners before you had uncontrolled bowel movements and relieved themselves in the straw, and you had to sleep on it. Getting in line for another potato while hoping*

and praying that there will be enough for seconds. Trying to remove the lice from your bodies and clothing.

Hiding your food so that the mice and rats could not eat it. Thinking of your family back home and feeling sorry for yourself. Feeling completely alone and helpless but thanking God that you were still alive.

I kept a partial diary by writing on cigarette paper and made notes of certain events. The following are some of the events that I did not record in my diary.

We were between Hamburg and Berlin and the Germans were rushing us in order to get us through this sector. The 8th Air Force bombed targets in Danneberg after we had walked through the town. Our group of POWs had been walking all day and the time was approaching 10 P.M. The sky suddenly lit up from flares that RAF planes were dropping. Then all hell broke loose as German night fighters roared overhead with their machine guns blazing away. We all hit the deck and did not move for several hours.

Another time when we were walking, a group of P-51s flew over and started dive-bombing oil storage tanks that were several miles away. Every time a tank blew we had to hit the deck again since the force from the concussion was tremendous.

As the days and weeks dragged on, Germany's infrastructure began to collapse to the point the POWs and guards had to forage food. On rest breaks, we would filter out in the farm fields and extract kohlrabies from the earthen mounds that housed them. The German farmers stored their crop of kohlrabies in these mounds for the winter. We would eat them raw since everyone was so hungry. The German guards were out of touch with their command structure for days and weeks at a time.

On another occasion, we were walking down a road that ran parallel to a railroad track. A train that was approaching suddenly stopped and the engineers jumped out of the cab and started running for all they were worth. Out of nowhere appeared a P-47 that proceeded to strafe the train. All of the Kriegies and guards jumped into the ditch next to road for cover. Suddenly there was a very loud explosion as the engine blew up.

The following event happened on a farm next to ours, where POWs had spent the night. The front lines were not too far away, since we could see the artillery flashes at night. The next morning, when the Kriegies fell out to start their day's walk, some of them buried themselves in the hay. The Germans turned the dogs loose on them. Still, that did not get all of them out so, the Germans set the barn on fire. They shot the Kriegies as they came running out.

Another time we walked by a German Luftwaffe airbase. This one housed the Me-262, which was a jet fighter. I had heard about the Germans having jet fighters but had never encountered any in combat. It seemed very strange to observe two of them taking off since they had no propellers.

Then there was the day that our group met a German SS Panzer Division that had pulled off the side of the road. Our guards warned us not to say or do anything but just walk past them very quietly and quickly. The guard said that they might shoot us if they found out we were Air Force POWs. I think that the guards were worried, even though they were in the German Wehrmacht. The SS wore black uniforms with silver piping and a skull and cross bones insignia on their lapels.

Towards the end of our long walk, our guards were changed to World War I veterans. They were not too happy about having to be in the army again and they seemed to be friendlier. Some of the anti-aircraft gun batteries that I saw fourteen to sixteen-year-old kids manned. Their uniforms were too large and their pants and gas mask canister dragged on the ground.

The flood of German refugees on the road was staggering, besides hindering our ability to walk. They were using every means available to move their personal possessions and flee the advancing armies. Then there was always the menace of strafing from Allied fighter planes thinking that we were German soldiers mixed in with the refugees. We began carrying a homemade Red Cross banner to help identify ourselves. We had not received any news reports since leaving Stalag Luft IV, so it was difficult to know just what was happening. The guards would occasionally give us an idea as to where the war stood.

Our biggest scare came on April 17, 1945 in the small town

Annaburg not far from where the Buchenwald Concentration camp, the Germans first and largest concentration camp, was located. Some of the guards told us they were taking us there but that never happened.

Following is a diary I kept of some of the events and places we stayed:

February

6:	Zarne Franz, 23K	Germans brought up soap kitchen. RAF bombed Stettin. Bedded down in open field. Anti-aircraft guns kept us awake most of the night. It snowed.
7:	Stolzenberg, 30K	More Volkssturm building bunkers.
8:	Kolsberg, 25K	Had blisters on right heel.
9:	Layover	Worked on blisters and rested.
10:	Greifenberg, 30K	Hard to walk.
11:	Kambs, 21K	Blisters turned into open sores.
12:	Layover	Worked on foot and made padding.
13:	Dobberphul, 20K	Foot much better.
14:	Pritter, 43K	
15:	Zirchow, 23K	Have got body lice.
16:	Murchin, 30K	Trying to survive. Dreaming of food, home.
17:	Medow, 16K	Got kohlrabi (German vegetable) to eat.
18:	Gultz, 28K	
19:	Ratzow, 18K	Walked two weeks now.
20:	Layover	Picked lice.
21:	Luplow, 23K	Everyone is picking lice off of their clothes.
22:	Layover	
23:	Layover	Wore hole in soles of shoes, had to be resoled.
24:	Layover	Shoes returned. Germans had hobnails put in soles.
25:	Layover	
26:	Layover	

27:	"	No news on march ending.
28:	"	Picked off lice.

March

1:	Layover	
2:	Karkow, 45K	POWs dropping out because of foot sores, dysentery, and pneumonia.
3:	Lebbin, 27K	Hobnails are making my ankles very sore.
4:	Layover	
5:	Zahren, 32K	Getting used to the hobnails.
6:	Strabendorf, 17K	Rumor some POWs have typhus.
7:	Layover	Have body lice really bad; they are eating me up.
8:	"	
9:	Zieslebbe, 14K	Came down with dysentery. Must watch so I don't start to dehydrate.
10:	Dutchow, 16K	Could not eat anything; dysentery really bad. Must keep walking. Ate some charcoal.
11:	Balow, 20K	Charcoal helped and I'm able to take some food. Still walking.
12:	Layover	Thank God!
13:	"	Spent the whole day picking lice off of clothing.
14:	"	
15:	Layover	
16:	"	
17:	"	
18:	Layover	
19:	Beckentin, 12K	Feeling much better.
20:	Brossegard, 17K	Crossed Elbe River.
21:	Danhoe, 31K	
22:	Bedenbuck, 17K	

23: Layover
24: Rhorstorf, 19K
25: Hohenbowstorf, 19K
26: Layover   "C" column split at Ebstorf, one section shipped to Fallingbostel. We were loaded on a train at Uelezen.
27: Uelzen, 12K
28:   Loaded on train fifty men in boxcar, 1/2 loaf bread and 1/2 pound of margarine for five men. Marshalling yard pretty well bombed out.
29:   On train: Everyone sweating out air attacks on train.
30:   Still on train.
31:   "   "

April

1: Arrived at Stalag XIA   Altengrabow, Easter Sunday. Camp has French and Indian Hindu POWs. Camp has a large area surrounded by double barbed wire fence about 12 ft. high. Eighteen of us in tent. Other POWs in large circus tent.
2: Layover
3: Layover   Bartered for meat with Indian POWs. Religion forbids them eating meat.
4: Layover   Everybody nursing foot sores, picking lice and cleaning up. Weather is much warmer.
5-8: Layover
9: Layover   Air-raid sirens go off. Germans fired rocket batteries.
10: Layover
11: Layover   Believe the Russians are getting closer.

12:	Layover	Evacuated camp by waking again. Many refugees on road.
13:	Lubitz-Belzig, 26K	German guard told us that Roosevelt died.
14:	Dannshore, 18K	Having trouble walking.
15:	Schonefeld, 20K	Can hear and see flashes from Russian artillery at night.
16:	Jessen, 18K	
17:	Annaburg, 19K	Stayed in old pottery factory; Germans declared Annaburg an open city.
18:	Layover	American front lines must be close. Can see artillery flashes.
19:	Layover	B-25s bombed fuel dump in the woods by factory. Everyone ran for cover. Found a crate with German bayonets in it. Put one in my bedroll.
20:	Layover	Evacuated factory.
21:	Dahllwbrl, 21K	
22:	Dommitzsch, 14K	American recon plane following us.
23:	Falkenberg, 4K	
24:	Krina, 36K	Seem to be walking in circles.
25:	Layover	Rumored that Germans were going to surrender. Tried to take a bath and almost froze. First time that I've had my clothes off in 2 1/2 months.
26:	Bitterfeld, 14K	Liberated at 4:45 P.M. by 104th Div. of the U.S. 1st Army.
27:	Halle	Went by truck. Was housed in barracks at a former German airbase.

May 7, 1945, Times Square, New York: Large overflowing crowds celebrate the news of Germany's unconditional surrender in World War II. *Courtesy: Associated Press/Tom Fitzsimmons*

FREEDOM

During the march, each agonizing day was a repeat of the previous day. As the days and weeks stretched into March and April, the weather began to improve. This helped to make life somewhat more bearable. What kept us going was that our liberation was not far off but the trick was to survive each day as Germany crumbled around us. Still the Germans stubbornly held on to the POW airmen. Even knowing that liberation might be at hand did not prevent some of my buddies from giving up. They just could not handle these kinds of conditions anymore. At times I thought that I had died and was in "hell." I had often thought of the remark made to me by a German officer after my capture about the war being over for me now. It was over in a sense but a different phase of it had begun for me.

We started our march with about seven thousand POWs from the

POWs celebrate their release upon news of their freedom. *Courtesy: Greg Hatton, b24.net*

Liberated American POWs wave as they leave their camp in 1945. *Courtesy: Joseph Eaton/U.S. Holocaust Memorial Museum*

entire camp and twenty-five hundred in "C" section, which was my section. We ended with less than two to three hundred in our section. The Red Army liberated some POWs while others ended up in other prison camps. A few of those the Red Army had liberated never were heard from again. We went through three groups of German guards, starting out with Luftwaffe guards but ending up with Wehrmacht guards who were World War I veterans.

The rumor was that the Germans knew the war was lost and that they were trying to move us into the mountains of Czechoslovakia. There, the most fanatical Nazis were going to make a last stand and hold us as hostages for bargaining with the Allies. Also rumored was that Hitler had ordered that we were to be eliminated.

Luckily, that never happened. American and Russian forces had met on the Elbe River, west and east of us, to cut us off. The German authorities negotiated with the Americans for our turnover and their surrender.

Los Angeles Examiner front page announcing Germany's surrender and end of the war.

On April 26, 1945, we walked the fourteen kilometers to the town of Bitterfeld. As we passed through some of the small communities, people were flying white flags from their homes. Many thought we were the occupying forces and were very upset when told otherwise. Others brought out containers of water for us. They sure showed a different attitude towards us then the prior weeks of marching with no water offered. According to the Potsdam Agreement, this area would fall into the Russian sector.

As we neared Bitterfeld, advance units of the 104th Division of the U.S. 1st Army met us. I cannot describe the elation and exuberance shown as we met these men. Our German guards laid down their guns and became POWs. Some of the members of the 104th helped us cross the Mulde River outside of Bitterfeld. With the permanent regular bridge destroyed, they constructed a temporary bridge in its place.

I was now free. I had survived this great 600-mile ordeal through one of Europe's worst winter in this century, existing on 770 calories a day. I had lost over sixty pounds and weighed in at 105 pounds. I had slept in barns, ate with the cattle, suffered dysentery, and malnutrition. I had open sores again on my feet and was having trouble walking when I awoke in the morning. I would have to take my legs and physically move them with my hands to get moving.

Freedom would mean food for my starved body. No more walking from ten to twenty miles a day in the bitter cold and living on food that would hardly fill a coffee cup. No more living like an animal. I would be able to bathe, get clean clothes, and rid myself of the 10 million lice living off me for the last two months. Now I would become a human being again. I believe the worst part of this march was the gnawing pangs of hunger that seemed to be present all the time. My dreams at night were of different tasty dishes of food. The happenings of the last three months finally caught up to many of us, for we sat down and cried. The events of the last three months had pushed many of my comrades to the end of human endurance.

May 7, 1945: General Alfred Jodl signs an unconditional surrender to the Allies at Reims, France, and officially end the war in Europe. *Courtesy: U.S. National Archives and Records Administration*

Donald H. Jones, a fellow Kriegi, wrote the following about our liberation:

> We came upon many soldiers of the 104th. The guards then handed over their rifles. We hugged, shook hands, tears were flowing and my personal feeling was that, in an instant, I passed from death unto life. Freedom was something I didn't ever want to lose again in my lifetime.

Four days later, on April 30, 1945, Hitler committed suicide in his bunker in Berlin.

While at the German air base in Halle, we picked up a Russian POW. I do not remember how we acquired him, but he did not want to go to sign an agreement with Stalin that all liberated Russian POWs returned to Russian control. On May 7, 1945, Germany had surrendered unconditionally. The war in Europe was then officially over.

A week after arriving at the Halle air base, I ran into a reporter named John Thompson for the *Chicago Tribune*, who was looking for freed POWs from the Chicago area. Of course, I told him that I was from northern Indiana. He took my name, told me he was doing an article on POWs and said that he would see that my name got into his article.

Meanwhile, back home, my family had not heard from me for well over three months. The Red Cross had advised families of POWs that many POWs were being evacuated from their prison camps by forced marches because of the advancing Allied armies. Since it was winter, they should expect some bad news because of these marches.

My father had passed away while I was in the service, so his brother, whose name is also John, was appointed to look after my affairs while I was gone. Uncle John was a foreman and worked in northern Indiana. On May 5, 1945, two days after I had met the reporter, one of my uncle's workers came running into mill office clutching a copy of the *Chicago Tribune* yelling, "John! John! Your nephew has been freed. He is alive."

Someone brought out bottle of wine and they all toasted my freedom. Nothing much was accomplished for the war effort that day.

June 18, 1945: General Dwight D. Eisenhower with General George C. Marshall waves to spectators at the airport on his return to Washington. *Courtesy: U.S. National Archives and Records Administration*

While at Camp Lucky Strike, I ran into a high school mate of mine, Seymour Lichtenfeld. He had served in the 106th Division and the Germans had captured him during the Battle of the Bulge. He was suffering from the effects of having his feet frostbitten.

Seymour and I were walking down a dirt road in the camp one afternoon when, all of a sudden, I saw a staff coming. I noticed that it had a five-star flag flying on the fender. As the car pulled up and it came to a halt, the driver jumped out and opened the rear door. Out stepped before our very eyes General Dwight D. Eisenhower, better known as "Ike."

August 8, 1945, "V-J Day": Back in Gary, Indiana with Seymour Lichtenfeld. *Courtesy: Seymour Lichtenfeld*

He asked what state we were from to which we both answered, "Indiana." He wanted to know what POW camp had held as prisoners. We told him.

After this, he said, "Soldiers, we are going to get you home as soon as we can." Then he turned around and jumped back in the staff car and left. We stood there with our mouths hanging open and just looked at each other.

Top: The USS Admiral W.S. Benson brings returning American servicemen back home. *Courtesy: NavSource Online Service Ship Photo Archive Bottom:* Servicemen cheer as the ship docks. *Courtesy: Associated Press*

GOING HOME

June 10, 1945, I boarded the Navy transport ship, USS Admiral W.S. Benson, previously known as AP-120 after being commissioned into service, at the French seaport of La Harve. After spending over a month at Camp Lucky Strike I was finally going home. "Home" what a great four-letter word. We were finally at a point that many of us had dreamed about during our imprisonment. Our saying that "You could plan all day and dream all night long, but the next morning, you still woke up in the same place" was about to change. They gave me a top bunk in one of the troop compartments. The only problem was I had to negotiate around some steam pipes to get in it. After leaving the harbor everyone went topside. We were all very excited.

They told us that we would not be part of any convoy and were going to try to make it to New York in three-and-a-half days. Off the southern

View of the USS Lafayette in tow and Hudson River piers with the New York skyline in the background, circa 1945. *Courtesy: U.S. Navy/New York Historical Society*

coast of England, they spotted a mine floating in the water. We circled it for thirty minutes with the marines on board shooting at it with a forty-millimeter deck gun. They had an awful time trying to hit it. A marine sharpshooter finally sunk it by shooting at it with his carbine.

Lashed to the deck were cases and cases of K-rations. By the time we reached New York, every outside case had holes punched in them. These boxes represented food and this boatload of ex-Kriegies was hungry. After all, we were only getting two meals a day on board the ship.

The evening before we approached the U.S. coast, the captain slowed the ship down to enter New York harbor the next morning, June 12, 1945. No one could sleep that night. Early the next morning everyone was on deck. As we approached the harbor, fireboats horns greeted us as we passed the Statue of Liberty upon entering New York harbor. The noise was terrific. Cars lined the Hudson River causeway tooting their car horns and people crowded along the bank waiving anything and everything they could leave. I do not think I can describe my homecoming except there were many kisses, hugs and tears.

After my leave, I was to report to Miami for possible reassignment to the Pacific Theater. The atomic bomb dropped on Japan in August while I was on leave. Following this event, Japan surrendered so the war officially ended. My shipping orders changed; they gave me another fifteen days' leave. Eventually, they sent me to an air base in San Antonio, Texas, where, on October 25, 1945, they discharged me from the service. I was now old enough to vote in the next election and to legally buy and drink an alcoholic beverage. I was twenty-one.

Top: Hitler and propaganda minister Joseph Goebbels during a walk at Hitler's mountain retreat, the Berghof, on April 12, 1943. *Courtesy: German Federal Archives*
Bottom: Hitler's Berghof heavily damaged after a RAF bombing raid on April 25, 1945. *Courtesy: Royal Air Force*

EPILOGUE

On February 13, 1945, and continuing into the next day, eight hundred and seventy British bombers, followed by five hundred American bombers, dropped tons of incendiary bombs on the city of Dresden, a transportation hub. This caused a firestorm that reached temperatures of one thousand degrees and devastated eleven square miles of the city. When the news of more than 135,000 German civilian causality figures reached Berlin, Goebbels advised Hitler to order the execution of all POW airmen. By doing this he felt that it would ensure that the Allies would not carry out another one of these attacks again. Hitler never took his advice.

In the last two months of the war at German Supreme Headquarters, Hitler was enraged by a report that German troops were still protecting downed airmen. He issued an order that all Allied airmen still under

A view from the town hall of Dresden, Germany's Old Town destroyed after the allied bombings between February 13-15, 1945 after 3,600 aircraft dropped more than 3,900 tons of high-explosives *Courtesy: Walter Hahn/AFP/Getty Images*

German control was to be turned over to the SS for elimination. He was tired of contending with these people. At the time of Hitler's order, Germany was in such state of final collapse that they never carried out his orders. The following is from the diary of Joseph Goebbels:

March 4, 1945

The Fuhrer is violently opposed to any steps being taken to assist Anglo-American Prisoners of War now in the process of being transferred from the East to the neighborhood of Berlin. There are some 78,000 of them and they no longer can be properly fed; they are riddled with lice and many of them are suffering from dysentery. Under present circumstances, there is little one can do for them.

At a Red Cross meeting in New York on February 21, 1945, Richard

F. Allen, vice-chairman of the American Red Cross, said that American prisoners of war in Germany were being marched deeper into the Third Reich. They were being marched through chilling temperatures as low as 30 degrees below zero without proper clothing. Those who had someone in a German prison camp must be ready for bad news.

Frank L. DeSilva of Seattle, Washington, a former POW from Stalag Luft IV, gave this account:

> We started walking February 6, 1945. The camp was at 54 degrees north latitude. We walked until May 4, 1945. We were actually liberated on May 2, 1945, but the government didn't get the word for a few days. Seven thousand men started from Stalag Luft IV. When they built a muster roll call three months later at Camp Lucky Strike, only 5,500 men answered the roll. I helped to bury many of them.

Later, unofficial reports indicated that about four thousand POWs survived this ordeal. Amoebic dysentery, pneumonia, other diseases and malnutrition had taken their ghastly toll. The prolonged malnutrition and abuse during captivity left many POWs pathetically drained of their energy and stamina. Some never regained their health and this shortened their life. In the post-War Era, a significant number were plagued with gastro-intestinal disorders and psychoneurosis. The after effects of the conditions that they endured would go on for many years. The majority ultimately regained their health and went on to live productive lives.

General Eisenhower made this statement concerning the fiendish practices of the Nazis, after his troops had rescued POWs from their German camps on the European front. General MacArthur witnessed the same perverted practices to POWs by the Japs. Eisenhower said, "The POWs lived under the most inhumane conditions. Prisoners of war were taken on death marches, transported in roofless cattle cars unprotected against the weather, or in sealed up freight cars devoid of air and light. In

Original drawing of Adolph Hitler by the author. *Courtesy: John L. Lenburg Collection*

Booklet celebrating the end of World War II and Hitler's reign. *Courtesy: Indiana Historical Society*

either case he captives did not have sanitary facilities, and many died during their journeys."

Railroad transportation problems grew graver in Germany and there were not enough trains to carry the captives, forcing the POWs to march hundreds of miles on foot in all kinds of weather, prodded by their captors' bayonets and bullets. Anyone who faltered were left to die by the roadside; the rest were beaten, kicked, bayoneted, and shot to death by their captors, in a most callous disregard of the Geneva. The advancing armies found many of the prisoners of war of the Allied Armies in a pitiable state of health. It was common for a man to lose as much as forty pounds in three months of imprisonment. Such loses resulted from an insufficient diet supplemented by a deliberate theft and retention of Red Cross prisoners of war food packages. Many of the packages were not stolen and consumed, but they were apparently withheld for no reason whatsoever except sheer cruelty. Three months after the Germans surrendered, nine million of these packages, stored away by the Germans, were found.

Being a prisoner of war or a prisoner of some foreign government is not like being a convict in prison in the United States. As a POW, you are not treated as a human being; you have no personal feelings and you are more or less a machine that is cut on or off by the military. They have no respect for you. You come and go at their request; your food, clothing and medical attention is as they see fit. Being a prisoner of war, you have plenty of time to think of the great country you are from and the personal freedoms that you enjoy. You take so much for granted.

Memories I will never forget of my imprisonment during the march:

> Trying to turn over on my side to sleep in the barn and there was no place to turn.

> Wishing I had a Red Cross parcel for every time I was told that I was going to be shot.

> The feeling of being completely helpless when our fighter planes made passes over our column.

Hope the rumors were true, that the Red Cross parcels were at the next village.

Getting back in line for another potato or soup ration, after burning your month by gulping down your first ration and praying there was enough for seconds.

Trying to remove the lice from my body and clothing.

Hiding any bits of food so the rats and mice would not eat them.

Thinking of my family back home and wishing I were there with them.

Going to sleep dreaming of food such as steak, eggs, and pancakes.

Feeling completely alone and helpless and sorry for myself.

Thanking God, I was still alive.

Happiness I felt each time I received my meager portion from a Red Cross parcel.

Elation, happiness, thankfulness and confusion when I was liberated.

The war crimes case against officers and guards of Luft IV consisted of fifty-eight volumes of testimony, forty-four that were statements from liberated prisoners. In its final form, *Reg. #1648 - United States against Germany - case #1*, named fourteen officers and guards as perpetrators of crimes against prisoners during the evacuation of Luft VI and afterwards at Luft IV.

In 1946, special U.S. agents were sent to confirm the identities Pichardt, Bombach, and Fahnert after their capture. They showed photos of Pichardt, Bombach, Fahnert, and camp staff personnel to a number of POWs who had witnessed the shootings of Sergeants Walker and Nies. The Judge Advocate was not able to obtain very many German records. The Germans had buried all of the American files before evacuating camp. No "paper link" ever established the accused to these events. Consequently, they were not brought to trial. The court also held that obeying illegal orders could not be a basis for a charge. Therefore, I believe justice was not served here.

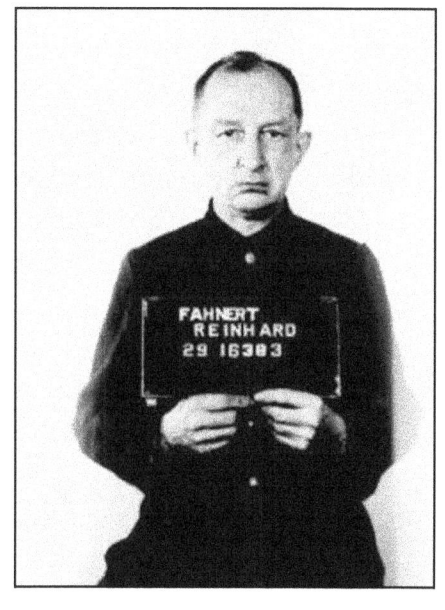

Captured German war criminal Reinhard Fahnert in 1946.

Previously, in April 1945, Big Stoop (Schmidt), one of the most notorious perpetrators of war crimes against POWs, was located in the Moosburg area dead with his head severed. One theory was that one of the Kriegies might have got to him. Maybe in some way there was some kind of retribution.

Erf, Barrowclif, Brown and Hendricks went to an officer's POW camp, Stalag Luft III, at Sagan. In January 1945, due to the advancing American Armies, the Germans evacuated Stalag III. The Germans walked these POWs to Spremburg, about one hundred twenty kilometers where they split them into two groups. One group was loaded into boxcars. Barrowcliff went on to Moosburg, Stalag 13D and Mike to Nurenburg, Stalag 7A. Due to the extreme cold weather, Mike Brown's feet were frostbitten. When the American Army got close to Nurenburg, the Germans manacled Mike and the other POWs, and then walked them to Mooseburg, a town that lies on the river Isar in Bavaria, Germany. Patton's Army liberated him, Barrowcliff, Erf, and Hendricks in April.

Leonard "Pappy" Bernhardt spent his entire confinement in the hospital in Budapest until liberated by the Russian Army. Afterwards he spent many years going through plastic surgeries to reconstruct his face. Ralph Wheeler is buried at a military cemetery in Lorraine, France, while a simple marker memorializes Martin Troy, whose body they never recovered, at the military cemetery in Florence, Italy. It is believed that Rube Waits, Jr. is buried in his hometown of Atlanta, Georgia.[5]

In an ironic twist of fate, six enlisted men from the 460th met their death and two injured themselves here in the United States after surviving their missions in Europe. Having completed their fifty missions, the crews from the 12th and 15th Air Force returned to the United States. After their thirty-day leaves were up, they gathered at Fort Sheridan, Illinois, to ship out to sunny and tropical Miami, Florida. On September 13, 1944 they boarded the train, the Dixie Flyer, for the trip to Miami. In the early morning hours of September 14, 1944, the Dixie Flyer slammed into the rear of an express mail train near Terre Haute, Indiana.

I have donated to the POW Museum in Andersonville, Georgia, the following items: a copy of this book, pieces of 20-millimeter fragments

Newspaper headline and story about the fatal train accident that killed returning vets.

---

[5] *In 2007, a U.S. team excavated Troy's remains from the crash site of the 460th Bomb Group's B-24 Liberator outside of Budapest. A year later he was laid to rest at Arlington Cemetery.*

taken from my face, a bible, a hymnbook and a video interview relating some my experiences as a German POW.

On September 4, 1992, the Polish Government erected a memorial at the former Stalag Luft IV campsite in the former Gross Tychów, Poland to honor the ten thousand American, Canadian, British, French, Beglian, Australian, Polish, Russian and other airmen kept as prisoners of war. It has the following inscription in English and Polish:

> *– This is the site of Stalag Luft IV*
> *– A prisoner of war camp for 10,000*
> *Allied airmen from the United States*
> *and the British Commonwealth*
> *– From May 1944 until their forced*
> *evacuation in February 1945.*

PART II:
CREW MEMBERS

ALAN BARROWCLIFF

The following is an account of events of June 30th given by our co-pilot Alan Barrowcliff.

June 30, 1944

That day, we were in the high box of the group flying on the right-wing position of Capt. John "Ham" White. He was supposed to be flying at an air speed of 155 mph; however, he was having engine trouble. Our speed showed much less and consequently we kept stalling and had a hard time trying to stay in position.

Our plane (B-24H #41-29291) had a large blue

"V" (for Victor) painted on each side of the fuselage near the tail section. Each plane in the squadron had a different letter of the alphabet and each of the four squadrons in the group had a different color. The letters were used for identification when contacting each other over the radio.

Approaching Hungary, we encountered many large white, billowy clouds. So many, that it was hard to keep the rest of the boxes of the group in sight. At 17,500 feet near Lake Balaton, Hungary, we were attacked by twin-engine Me-410s. They stayed just above the clouds and shot at us with 20-millimeter cannons, but we could not see them or reach them effectively with our 50-caliber machine guns. Later, when they started to make passes over us from the rear and going forward, I saw tracers from Jack Nagel's nose-turret hitting the port engine of one of the fighters and putting the engine on fire.

One of the engines on our left wing caught fire and brown smoke started coming up from under the instrument panel. The glass tubes of the gas gauges at the rear of the flight deck had been hit and were burning. It all sounded like hail on a tin roof when we were getting hit.

Strings of 50-caliber ammunition were hanging down onto the flight deck from John Lenburg's upper turret, then the fire started exploding the rounds of ammunition adding to the noise.

By this time, we had dived down to 12,000 feet to try and escape the German fighter planes and leveled off to let everyone jump. While Erf held the plane level, I pulled the handle, which was supposed to open the four bomb-bay doors in an emergency. On the second pull, all of the bombs should be released. The forward port door was the only one that opened and released the four 500-pound bombs in that bay.

John Lenburg had gotten out sometime before I tried. On my first attempt to get off the flight deck through the small doorway leading to the bomb bay, I was singed by the flames from the gas gauges. Then I opened the top hatch that was behind the pilot's seat and in front of the top turret. After climbing up and looking out, it did not seem that I would be able to jump from there without either hitting the propellers of the two inboard engines or, if I got past them, I would probably hit the rudders and knock myself out or get killed instantly. From this spot, I could see the flames coming out of the back of the plane and that the left wing was on fire.

So, I dropped back onto the flight deck. I felt something hit my ankle. Seeing a small rip on my flying boot and later a nick on my G.I. shoe, I figured that it must have been a bullet from the exploding rounds on the flight deck.

I realized then that there was no alternative and that I would have to dive through the flames to get to the doorway that led to the bomb bay.

As I dove through the flames, I saw something burning on my right shoulder and figured that it was my parachute strap that had caught fire. The catwalk of the bomb bay was three or four feet lower than the flight deck level; so, I had to reach down and pull myself through the doorway and then twist my body to the right in order to get through the bomb-bay opening.

By this time, I did not think that I had a parachute anymore on my back since I had seen something burning on my shoulder. But after being partially burned on my hands and face, I knew that I did not want to be burned to death and would jump without a parachute.

The next thing that I was conscious of was a sharp pain in my right groin area. It was a parachute strap pulling up tight that caused the pain when the parachute opened. It was a beautiful sight to look up and see a white canopy all in one piece, fully opened, and not on fire. What had burned on my shoulder was the first-aid kit that was tied to the straps.

I have no recollection of counting to ten before pulling the ripcord to open the parachute, and I must have dropped the metal D-ring right away. The plane was not very far above me, so I must not have waited too long before pulling the ripcord. I do not remember actually leaving the plane; I could have passed out from the pain in my back while twisting to pull myself out. My watch showed 10 A.M.

When I looked down, I was floating right over the middle of Lake Balaton. However, the wind was blowing me towards the north shore. There was no sensation of falling at that height. After all the noise of the cannon shells and exploding ammunition hitting the plane, it was relatively quiet now.

Facing towards the north, I could see four chutes in a row below me in the distance to my right; above me, on my left, I could see one more, which had to be Erfeldt's since he would have been the last one to jump. Later, I found that the four chutes belonged to Nagle, Hendricks, Brown and Lenburg; Bernhardt had jumped earlier. These were the only men to survive; Wheeler, Waits, and Troy died in the plane.

There were several burning wrecks on the ground and our plane made a slow turn to the right. It then crashed into the hillside many miles away. The weight of the bombs still on the right side of the plane probably made it turn in that direction. A German Me-109 made a pass close to me below my feet but continued towards the west end of the lake.

As I got nearer to the ground, it seemed to come up with a rush and it was easy to tell of the sensation of falling. I could see that I was going to land in a wooded hillside area, so I pulled on the shroud lines to try and guide myself towards an open field. There was a small cottage at the edge of it with several people standing in the doorway watching me come down. So much of the canopy collapsed, spilling out a lot of air. It made me afraid to try it again for fear I might make it collapse altogether. So I just watched the trees rush up quickly towards me. I remembered from the instructions class at Primary Flying School that I should cross my legs and cross my arms in front of my face if you come down in the trees.

The next thing that I knew, my feet just touched the ground and the parachute was hanging on the top of the trees. It was impossible to pull it down to hide it amongst the bushes, which were about three-feet tall. I took off my sheepskin boots and tossed them under a bush after unbuckling my straps to get out of the chute harness. Then I ran several hundred yards through the underbrush to get away from the parachute area, since it would not be hard to spot and the people would search for me right there

I laid low in the bushes on the hillside and took stock of my condition. I used the salve from my escape kit to put on my burns. The tips of my fingers and the tip of my nose were blistered. The back of my right hand had third degree burns and each side of my face at the sideburns area.

Most of the right leg of my flight coveralls was burned off and near the ankle area of the left leg was, too. I do not know about the back area of the clothing. It was very hot since the sun was shining and I was dressed for high altitude flying. I had on cotton underwear, khaki slacks under "pink" slacks, with a wool, scotch plaid flannel shirt over a khaki shirt, a pair of flight coveralls, and a leather A-2 flying jacket on top of everything. I had pulled off my sheepskin-lined helmet at the same time that I removed my boots. The helmet had covered my ears and the hair on my head thus keeping the flames from burning me there.

My hunting knife was the only weapon that I had since I had left my 45-calibre automatic in the plane next to my seat. I always took it with me on the plane in case we ever came down in the water and it would be useful to break the windows if it was necessary to get out through them.

I looked at my escape maps and buried the ones of Spain, France, and Germany. It looked like I would have to head westward around the end of Lake Balaton and walk towards Yugoslavia. The lake is about 50 miles long, running northeast to southwest, and about seven miles wide where we bailed out of the plane. The nearest town was Tapolca, which is a third of the way from the southwest end of the lake.

It did not seem too long before I heard voices and people calling each other. I laid low in the leaves and bushes and could only see shadowy shapes passing about a hundred feet from me. It was not possible to tell whether they were in uniforms or not, since I was looking uphill towards the sunlight and it was fairly dim in the woods. If they were friendly civilians, I did not think that they would have made so much noise. Later I found out that they were Hungarian soldiers looking for me after finding the parachute.

I ate a piece of candy that was included in the escape kit, but I had nothing to drink and it was hot and uncomfortable with all of the clothes that I was wearing. Because of my burned hands, I could not remove the leather jacket, which was the hottest item. I also had on two pairs of socks under my G.I. shoes.

At 14:00 (2:00 P.M.), I had not heard any voices for a long time, so I decided to start walking. Heading west over the top of the hill through the wooded part, I came to an area that was partially bare of bushes and trees, just waist-high grass. I climbed a small tree in order to look around to see which the best way would be to get down the opposite side of the hill. Many miles away, I could see the end of the lake.

After climbing down out of the tree, I started to walk a narrow footpath leading downhill through the tall grass and weeds. I had not walked too many steps along the path before I saw a peasant walking towards me with a big smile on his face. From his appearance, he had to be over 60 years of age, and when I talked to him in English or German, I did not have any success in holding any conversation. However, I motioned towards the cigarettes in my breast pocket and, through sign language, I indicated for him to reach for them, to take one for himself and to place one in my mouth.

Since there were no signs of hostility from him, I was hoping that I was lucky and had met someone that might help me with the underground people in that country. He motioned for me to follow him and we started down the hill following the path.

At the bottom of the hill were four Hungarian soldiers carrying rifles waiting for us. Nearby were several peasants standing outside of their cottages watching us. The soldiers seemed friendly and I could converse with one of them in German. He said that he thought the English, Americans, and Hungarians were gentlemen, but the Germans were not good. The soldiers were from a Hungarian pilots' rest camp, which was nearby where they picked me up. A lot of the shooting that I had heard earlier was from a rifle range at the camp.

We started through the fields of grain with two of them in front of me and two behind. I was told that they were taking me to an airfield that was not too far away. We would pass peasants that were out in the fields cutting and stacking the grain. I could see them waving their pitchforks and scythes while hollering something. It was explained to me that they thought that anyone in flying clothes was a "Terror-flieger" (fighter pilot) that had been shooting and strafing them while they worked in the fields. It seems that our fighter pilots, when returning from escorting our bombers to a target, used to dive down and shoot at them while they were harvesting the grain or doing work in the fields. Consequently, they wanted to harm anyone who was captured. Fortunately, I had the soldiers for protection.

After a while, we came to a group of cottages; the soldiers took me inside one of them. We sat at a table and an old woman and a younger one gave me red cherries to eat and a shot of clear liquid, which was very strong. I presume that it may have been vodka since I had never had any before that time.

After a short rest, we continued walking towards the airfield that was still several miles away. The airfield turned out to be a training field for Hungarian cadets. Out on the ramp were parked over a dozen American basic trainers with blue fuselages and yellow wings, but they had German crosses painted on the wings and fuselages. The planes had been liberated from France when it surrendered to the Germans before the United States entered the war.

There were also two P-43 Lancers on the flight line, and several German planes, but I do not remember the types. As we approached the buildings, a plane took off and made a sharp 90-degree turn; it was a Fieseler Fi 156 Storch. I had never seen a plane make such a sharp turn.

I was taken into their hospital, or first-aid area, where they removed my clothing for me and then dressed my burns with ointment and bandages. They only allowed me to keep my underwear, khaki shirt, trousers, shoes and socks. They kept my watch, insignia, etc.

Then I was placed in a cell that had for beds, a shelf of boards about two feet off of the floor and sloped down at a slight angle. It was wide enough for three persons and had three "pillows" which were triangular-shaped blocks of wood. My roommates were Jack Nagle and a fellow from another crew. There were also some men in other cells who had been shot down.

LEONARD BERNHARDT

On June 30, Leonard "Pappy" Bernhardt was injured so badly in the attack by German fighters that he was not sent to a prison camp but was held at the Royal Hungarian Hospital No.11 in Budapest. He was liberated by the Russian Army when they captured Budapest in February 1945. "Pappy" was returned to the 15th Air Force in Italy.

The following is a copy of the report that published after he was interviewed by an American officer on March 10, 1945. When the word, "source," is used in the report, it is referring to Bernhardt.

The second article that accompanies the published report is a copy of a newspaper article that appeared in the *Boston Globe* on May 20, 1945.

SECRET   USA/SKP/667:
              : SECRET   :
              :AUTH: CC, 15 AF :
              : Init._____:

HEADQUARTERS FIFTEENTH AIR FORCE : 10 MARCH 1945 :
APO 520   U. S. ARMY :::::::::::::::::::::::::::::::::::::

B-JON-ag
10 March 1945

Escape Statement

1. Bernhardt, Leonard, S/Sgt., 31307604, 760th B. Sqdn. 460th B. G.
   Born - 15 October 1914   Enlisted - 1 April 1943
   Home address - 193 London St., East Boston, Mass.
   Peacetime Profession - Advertising
   MIA - 30 June 1944   RTD - 9 March 1945
   Number of Missions - 40
   Number of Sorties - 32

2. On the way to the target, box of source's A/C [aircraft] was separated from the rest of the formation by bad weather. They were attacked by Me-110s (source saw 12, but there may have been more, as visibility was poor). Source knows four A/C out of total of five were shot down, out of a total of five. In case source's A/C, source was wounded, interphone was shot out, tail gunner got one Me-110, and No. 2 engine caught on fire, in that order. The entire section of wing caught, then source saw fire in the bomb bay. Source believes the bomb load had been salvoed.

Source ripped off flak suit and went back to get tail gunner out. On his way he tried to open the camera hatch, but it would not open. It had been in good condition at time of takeoff, but source thinks it had been welded shut by heat or jammed by fragments. Source got back to the rear and found the gunner dead and turret damaged by gunfire. When he turned back in the tail of the A/C [aircraft], flames were right in his face.

Source fought his way back to the waist and got his parachute that had been wrapped in a flak suit. The other waist gunner was standing there in the flames, seemingly paralyzed by fear, wearing his flak suit. Source motioned to him to get his chute but got no response. Source started out the waist window, but got stuck, being held partly by centrifugal force. The other waist gunner then pushed him out. Source was semi-conscious, but believes he opened his chute at once. Source did not see aircraft after getting out.

3. Hungarian Hospital at Veszprem (4707 N – 1768 E)
Hungarian Royal Military Hospital No. 11 at Budapest

4. On the way down, source was buzzed twice by a Me-110. Each time the A/C came straight at source, who dodged by spilling air from his chute (source was told later by some Hungarian pilots that there had been several such cases, where a pilot clipped shroud lines on a chute as a form of sport). Source landed in a wheat field at the north west corner of Lake Balaton.

He was found by civilians who beat him with scythes, pitchforks, hoes, etc. Hungarian gendarmerie, or home guards, arrived and rescued him from the civilians, and, in spite of his wounded leg, marched him down to the police station. Source was refused first aid. He was interrogated as to his target and slapped when he refused to answer this and other questions. A few minutes later, others from source's and other crews were taken in a German truck, with German guards, to a civilian hospital at Veszprem. All of source's valuables had been taken by the gendarmerie, who laughed when he asked for a receipt.

At the hospital, all of the wounded men were hospitalized and treated well, while the rest of the men were taken off to some prison camp. Source stayed in the hospital seven or eight days and believes they did the best for him. Suffering severe burns and wounds, source and other men were transferred by streetcar and train, to the Royal Hungarian Hospital No. 11 Gombos Gyula Utca (near the Vérmező) in Budapest, where he remained as prisoner patient until liberated on 7 February 1945 by the Russian Army.

Arriving at the hospital the first week in July, the source, who is Jewish himself, heard that Jews from all over Hungary were being herded into Budapest, some of them from as far away as Transylvania. In September, visitors to the hospital told source that ghettos had been established all over Pest {Hungary], walls had been constructed around them, and the Jews had been left in them to starve. Many were found shot in groups of two or three in doorways. Many were pressed into labor battalions, and the wounded from these battalions were brought to the hospital. More were segregated, being treated by Jewish doctors, who were allowed to treat non-Jewish patients. The Christian doctors were not allowed to treat Jewish patients.

The source's underground contact, said that he had seen 300 Jews herded to the Danube and shot by SS men, then weighted and thrown into the Danube. On December 10, the source heard from Tocay that thousands of bodies with the Jewish yellow star were breaking loose from their moorings and floating into the Danube.

Attached herewith are pertinent excerpts from a letter carried by the source, written about the middle of February 1945 by a Jewish family of Budapest to relatives in New York.

Extracts from letter

    FROM: Foro Chaura Optica Budapest
                  Ferenciek-tere 2
    TO: Mr. Martin Weiss, Refrigeration Service
                  Liberty Ave. 150
                  Jamaica, New York

To tell it in order, the cyclone caught me a year ago. With the German occupation, power was seized by the traitors of the country, the persecutors of the Jews. They chased us out of our villa on St.

Gellért; we had to leave everything within 24 hours. The store had to be closed, but wages for the personnel, rent, and taxes had to be paid under penalty; of internment. Evenings we were afraid to undress, for we were waiting in torture for the men of the Gestapo, to be taken to a place from where there is no return. We had to put on the yellow star, were allowed to live in designated houses, and could go to the streets only between 1 and 5 PM.

Our blackest turning point was October 15. Horthy's government failed and the Arrow Cross terror started. They withdrew every concession; they fenced the ghetto and closed in 80,000 Jews, without food, water, electricity, and gas. With cruel shrewdness they have searched out those hiding, and every night organized programs. About 20,000 men and women were handed over to the Germans. Off those we did not a thing. It seems they have disappeared forever.

After we were tortured by these things to the end of our endurance, the siege of Budapest started. And it lasted six weeks. We had to stretch our strength to the limit to beat the many privations. We kept the secret from each other that we were hungry. We told each other to carry on, that we have to see the end of it. And we lived to see the day. The Russians took the town and we were released from the ghetto, but outside we could only see a heap of ruins. Our store was robbed, our house as well. Small belongings were left here and there. But I do not complain. Thank God that all three of us escaped.

So far, I have only written about ourselves, but those from the provinces have their own tragedy. They were deported in closed cars, and we will only know after the peace treaty that is alive among them. For this the guilty parties will have to suffer; we talk about it, those who are left alive.

Three weeks ago, we came here to Kecskemét with Évi, for there is starvation in Budapest. Laci stayed in Budapest for the time being, trying to get to our valuables.

We try to recover and forget those experiences of hell. One thing is sure, that we cannot remain in Hungary, for if we have to start again, and it would be suicide in this country. We are asking you to attempt securing entry for us to America. There are no Legations here; we cannot take step without assistance. Perhaps it would be possible help of Alex Acél to Buenos Aires.

Medical Conditions During Siege of Budapest

Professor Fornát, an intestinal specialist from the University of Debrecen and a member of the Hospital No. 11 staff, told source last week of December during the siege of Budapest that the 4,000 German wounded were housed in Gellért tunnel, there were several cases of typhus.

At No. 11 Hospital itself, there were over 100 cases of paratyphoid among about 2,000 patients and refugees who spent the last five weeks of the siege in the first-floor air-raid shelter of the hospital. During this period, they had only one meal a day and undrinkable water. Medicine was very scarce.

Hospital sanitation even before the siege was poor. The staff was lackadaisical. The doctors were just out of school. Nurses and attendants handled bandages and instruments with bare hands and used the same instruments on various patients without sterilization, between times.

Source remained at the Royal Hungarian Hospital No. 11 until 7 February when again wounded (see Appendix E) and then by evacuation through a series of Russian hospitals, source was finally placed in Kecskemét Hospital, where he was contacted by members of the American Bomber crew that flew him out 9 March 1945.

Beginning on 25 December during the siege of Budapest, the Russians showered the city with plane-borne leaflets urging surrender. These were mostly in the form of articles by the Hungarian Generals Miklós and Veres.

The Russians were not firing on the Hungarian soldiers, in order to facilitate their desertion, and the Germans were providing German uniforms to the Hungarian soldiery. From Lt. Col. Pasztahy, the source learned that the German Commander had wired Berlin requesting to capitulate and was told to fight to the end.

The Royal Hungarian Hospital was the primary target because the Germans had placed guns around the hospital and had been using it as a communications center and as a point upon which to parachute supplies. On February 1, the battle line extended through the hospital grounds, which had suffered a total of about 1,000 hits. After the Germans were finally driven out, the Russians immediately began looting, and many were seen with jewelry and watches. They attempted to molest the nurses, several who ran to the source for protection because he was an American. Source assumed officer rank and ordered the soldiers away.

Russian Hospitals

Source was very well and efficiently treated in all the Russian hospitals through which he was sent. The Russian doctors, of whom about half were women, were older on the average than the Hungarian (doctors). There did not seem to be any shortage of nurses or medicine. Source was impressed by the speed and efficiency with which the Russian wounded was evacuated away from the front.

Impressment of Hungarian Labor

At Budapest, Dunaharaszti, and Kecskemét, where the source was moved and hospitalized by the Russians from 6 February to 9 March 1945, he observed that all types of Hungarians were impressed from the streets into labor battalions regardless of their occupations or personal affairs. This forced labor sometimes lasted 2 or 3 days, and usually they had to provide their own food.

Russian Requisition, and food, and equipment

Along the way from Budapest to Soroksár and Kecskemét, the Russians drove off cattle along the highways without payment to the farmers, and with no system of recording the requisitions. At Soroksár a rather poor peasant woman to who the source spoke in German (she was not a Swabian) said that the Russians had taken her meager livestock of ten cows, her pigs, and chickens, without payment. The general opinion of about twenty country persons to whom the source spoke, was that the land was being completely stripped of what was left of livestock that the Germans had not taken.

Because of this living on the land, the Russians were everywhere very well fed, eating meat three times a day, and having wine at every meal. The Hungarians themselves maintained that they had no wine. Russian officers spoke of this use of Hungarian food as a matter of course in the war, and that it was repayment to the Hungarians for the war.

The only good mobile equipment that source saw was American vehicles - Jeeps, Studebaker and Dodge trucks, some of which proved to have USA mark on the right side of the hood marked out. The Russian vehicles were antiquated wrecks.

Financial Intelligence

Source obtained thirty pengos for blue seal dollars in July; by October it was possible to get 120 for 1 dollar. During the siege 300 pengos were offer ed. In Kecskemét 300 to 400 for a dollar was the market value. Local people preferred the old Hungarian money to the Russian invasion notes.

J.C. RAMSEY
Capt., Air Corps, Interrogator.

ROBERT HAMILTON
Co. B. 2677th Regt.

5 Incls:

Incl. 1, Ltr 363.d
Incl. 2 - 5. Appendix B, D, E, F.

APPENDIX B

The Hungarian Underground

The source believes that there was no general organization of the underground in Budapest. Before the middle of November, when all the flyers except the source were evacuated to Germany, offers for aid were so immature that not one of the flyers at the hospital attempted to escape. After that time, the source and others at the hospital made contact with what appeared to be isolated groups specializing in escape (see Dr. Erno Vida, below) and in sabotage (see János Tócsy, below).

The escape groups specialized in certain forged documents and clothing necessary to impersonate Hungarian soldiers on furlough—furlough papers, soldiers' books, soldiers' caps and armbands (the entire uniform being unnecessary). In all, the source believes that the hospital groups aided 250 to 300 Hungarian officers to desert.

The ordinary "Igazolváy" form, filled out only on one side was still in use in December. Also, the "Urlaubachein" form. Also, the regular Christian birth certificate, which was forged to provide Jews with the means of Christian identity.

Underground Personnel

Dr. Erno Vida, eye specialist at Royal Hospital, Czechoslovak by birth; home address: Albert utca 84. In July, Dr. Vida listened to BBC and gave news to the boys. He was transferred to #10 hospital in August, but on returning in November he told source of his underground organization called "Makos," which appeared to be proficient in forging papers, arranging escape, and other underground activities. After November, Dr. Vida had prepared escape plans for deserting Hungarians and for flyers who might be coming in. He had all necessary papers and stamps, soldiers' caps and armbands (all that was necessary for impersonation of furloughed Hungarian soldiers).

In this activity he was aided by his brother, a sergeant deserted from the Hungarian Army. Dr. Vida also provided Christian birth certificates to the source for use of Eva Zabroy's Jewish friends. In December seven high members of Dr. Vida's organization were arrested and executed by the Gestapo. After Christmas, he did not appear at the hospital, and the source assumes that he had gone over to the Russians.

Lassle Eggenhofer, clerk in the eye clinic; home address: Budapest XII Cambos Gyulau 49/b III/8. Pro-Allied from the beginning but did not join Vida's group until November. Erased the source's name from the shipment list 18-19 December 1944. Left on Christmas to go to his family in Keztergom, which had been occupied by the Russians.

Dr. Sukavarahy, in charge of prisoners. Very pro-Allied; delayed the shipment of many flyers for months. However, during the siege of Budapest he marries a nurse of the Ear and Nose clinic who was, according to Vida and Eggenhofer, a Gestapo agent, member of the Arrow Cross. She is blond and wears thick glasses. Name unknown. On December 24 he went to her apartment in Budapest to live during the siege.

Countess Gabrielle Széchény, Austrian, head nurse; home address: Budapest XII Cambos Gyulau 49/b, Apt. III/1. Aided in plans to hide flyers in private homes in Budapest. Tried to prevent shipments by placing flyers in the Swedish Red Cross. Distrusted Count Craig but worked with him because he seemed sincere. Went to see the last group of flyers entrained for Austria at the risk of her life. Still at hospital at time source left.

James Tócsy, a lieutenant of the Hungarian Army, lived at the hospital as a patient, but was well. Member of an underground sabotage group of young Hungarian Army officers; asked source for military information which source secured from German wounded. Source provided Tócsy with Vida's forged documents. Lived in Pest [Hungary] during siege.

Eva Zabrey, X-ray department assistant; home address: Clasz Faser I. Gave source a letter from Pilot Officer R. Barrett, an escaped PW known as "Darky," head of an underground escape group, concerning plans for escape of Allied flyers, associated with Elizabeth Mike, but not with Dr. Vida. Went home during siege; bombed out and went elsewhere in Pest [Hungary].

Elizabeth Mike, X-ray technician; home address: Budapest III, Ker. Szépvölgyi - utca 26 FS. Gave source a letter from a Dutch officer "Dutchy," saying that Barrett had left and to plan escape thenceforth with him. (Ed. Note: Dutch officers hiding out in Budapest were outstanding escape experts in the underground, see H Report No. 18) Source replied to Dutchy through Mike, suggesting a contact with Dr. Vida. Miss Mike went home during siege.

Prof. A. Kettesy, chief of Eye Clinic at Hospital No. 11, director of Eye Clinic University of Debrecen, on staff of Royal {Hungarian] Hospital [No.11]. Gave source the run of the hospital grounds without guard after November 17, when he was assigned to the Eye Clinic. Prevented source from being sent to Germany with last group of flyers on November 17. Pro-Allied. Remained at hospital during siege.

Csady, Lt. Col. Commandant of the hospital. Very pro-Allied, helpful to the flyers.

Leichtner, or Leitna, Lt. Col., Vice-Commander of the hospital. Very pro-Allied, helpful to the flyers. Prevented the Germans entering the hospital with arms. Appealed German Commander not to fortify the hospital during the siege of Budapest; appeal was refused.

Kovassy, Gyula, candidate officer of the Hungarian Army, a deserter hidden as a patient at the hospital by Dr. Kettesy; home address: Debrecen, Egymalom utca 3. Appeared to be member of an underground group, and asked source for military information. After December 26, Kovassy said he evacuated his own family westward with the Germans to save them from the siege. Remained at hospital during siege.

Countess Lottie Keglevitch. Visited hospital, pro-Allied. Said she helped Hungarian officers hide in Budapest.

Lt. Col. Artura Pasztchy, a patient at the hospital, Commander of the Royal Guard; home address: Budapest, Attila Krt 4. Gave food to source and was good source of general information. Spoke English; was 15 years in Canada. Said he had given the order for the Palace Guards to fire on

Germans during Horthy's attempt to oust the Germans. Remained at hospital during siege.

Kiss, György, wine merchant, Okl Gasda E.N.V.T.U - Veseto; home address: Budapest IX Üllői - út 91, I EM 8. Visited source on Christmas day and asked him to hide at his home. Said that he was born in America.

Pro-Allied (Not underground)

Dr. Smelansky, Polish X-ray specialist. Elderly, pro-Allied captain in the Polish Army. Released from status as a POW to work on staff. Lived at hospital with family.

Russian Personalities in Kecskemet

Major N.D. Zigalkin, Commanding Officer of the Air Battalion at Kecskemét, is something of a legendary figure with the Russians. His power is apparently far greater than ordinarily associated with similar rank. Stories of his good treatment of Americans have been brought back by all the returning airmen. Major Zigalkin's battalion is completely self-sufficient with service troops and entertainment facilities for the soldiers. Zigalkin's civilian wife was with him in Kecskemét, as were the wives of many of the other higher-ranking Russian officers. The Major told source that when Vienna was taken it would be his next post. (Photograph attached).

Capt. Moore, a Russian whom the source met in Kecskemét, said that he was an American who had spent 25 years in Cleveland and returned to Russia just before the outbreak of the war for a visit. During the visit he was drafted into the Russian Army. Since he spoke excellent English he was helpful in interpreting the wants of the airmen to the Russians. Moore spoke quite derogatively of the Russian system and expressed the desire to escape from them back to the USA.

APPENDIX D

Pro-Nazi Personnel

Count Andre Orazagy, member of the Home Defense. Visited hospital and talked very pro-Allied. Had a van, papers, and clearance at his disposal, and made very elaborate plans for escape by air, but nothing of his ever materialized. He implied that the final evacuation of flyers on 17 November 1944 was a plan of escape, and the flyers thought up to the last minute they were escaping, when in reality they were being shipped to Germany.

Countess Huberta Széchény, volunteer nurse. Elderly woman, mother of pro-Allied countess Gabrielle Szecheny. At hospital during siege.

Lt. Col._____ (name unknown), Hungarian, old, a patient at the hospital. Remained after the siege of Budapest. During the siege this man kept insisting that the American airmen (the source) be shot.

Enemy Propaganda in Hungary

Germans spread propaganda that American bombers were dropping explosive fountain pens, pencils, dolls, toys, etc., over Hungarian towns. Hungarians reported that this caused beating and killing of American airmen. Hungarians were told, in newspapers and posters, that it was their duty to kill any American they saw coming down in a chute, as these men were coming down only to kill Hungarians. This held true even if men were bailing out of a burning aircraft, etc. Many men were in the hospital who landed safely.

The propaganda campaign against airmen had already been in effect a long time before the source's arrival in June 1944. As soon as he landed by parachute in a seriously burned condition, he was attacked by peasants with pitchforks and spades and almost beaten to death. He was saved only by the Hungarian military.

During his stay in Hospital No. 11, he followed the propaganda campaign closely, since as a newspaperman and victim he took a special interest in the subject. Atrocity stories concerning American airmen appeared regularly in the newspapers and magazines, such as *Pester Lloyd*, *Szikra*, and the German edition *Signal*.

These articles were all in the nature of press releases indicating a common source. The same themes were reiterated: Americans are barbarians, American flyers are a low type of humanity -- Jews, Indians, etc. Those stories often dealt with other aspects of propaganda, such as the post-war aims of the Allies to divide and despoil Germany. One common story was that each flyer received $1,000 per mission and $2,500 at the end of twenty-five missions and then go home to live a life of luxury. The "victory girl" campaign indicated that there were two girls allotted each ship for the crew's pleasure on returning to the airport. One of the posters complemented this theme by showing the girls waiting at the airport.

The source believes that almost every article singled out Jewish flyers for vilification. He also believes that few Jewish flyers escaped being killed on landing if they did not hide their identity as Jews.

At the time of the Cabinet change in July and continuing under Szálasi, the anti-air men campaign was relaxed except in the German magazines. This indicates, to the source that the campaign may have been instigated by the Arrow Cross Party (Nyilas) rather than by the Germans. In July, the story that airmen were dropping explosive toys and other articles was retracted in the Hungarian newspapers, and it was maintained that this was done by the Jews of Budapest.

According to the source, there was no anti-Negro campaign. A Negro flyer in the Royal [Hungarian] Hospital [No.11] was treated as well if not better than other Allied patients, being an object of curiosity.

The poster campaign, part of which was anti-airmen, was very widespread, the posters appearing on barns, in all public meeting places. When the Germans came in October, no new posters appeared, although the old ones remained. This also indicated, to the source, that the campaign was primarily instigated by Hungarian, not German, authorities. (Source brought one poster with him). After October, Hungarian periodicals no longer carried anti-Allied material, although pro-German articles appeared. German magazines continued the propaganda campaign until Budapest was taken. One article had pictures showing the excellent treatment of Allied airmen in German hospitals.

Effect of the anti-airman campaign on Hungarians:

Public opinion, including the middle and working classes and especially the peasantry, was seriously affected. Flyers were mistreated all over the country. Jews were usually killed immediately if they did not hide their identity. Since the source was in no condition to hide his dog tags marked "H," he was at first mistreated in the hospital by anti-Semitic nurses.

In an unidentified southern jail, one entire crew was stripped (probably to determine by circumcision whether or not they were Jewish), made to lie on the floor, and beaten with sticks by the Gendarmerie. Of this crew, Lt. Erickson, 2nd Lt. Lambert, and S/Sgt. Charles Robinson landed in Budapest hospital, where source learned of the matter.

The source believes that only professionals, the intelligentsia, and upper classes who were pro-German before the war disbelieved the propaganda campaign. Source believes that propaganda had a greater effect in the country than in the cities because of the lower level of intelligence and lack of independent judgment. As the Russians approached, however, the source observed and heard from others outside the hospital of the opinion becoming increasingly pro-Allied. Anti-German feeling became bitter as the Germans looted on their retreat. By the beginning of November, the Russians were welcomed.

APPENDIX E

Shipment of Allied Airmen to POW Camps in Germany

Source was the only American flyer left out of the last group of Allied flyers at the Royal Hungarian Hospital No. 11. The others, 45 in number, were shipped by rail to Germany on the 17th of November 1944. Between July 8th, when the source arrived, and November 17th about 300 Allied flyers, almost all of them Americans, were shipped from the hospital to Austria or Germany. More than 300 were shipped from the Budapest jail, where after capture they were usually kept five days and interrogated. Source estimates about 700 in all, mostly Americans, were sent from the city jail and this hospital.

Every week a van came from the city jail with flyers from there, and the Allied flyer hospital cases that were well enough to travel were taken away with them. Loads varied from 5 to 20 per week, although some weeks none were taken. It was fairly well established that they were taken to a railway station and went by train through Győr and Sopron to Vienna.

The name of the driver of the van is Szabó, an employee of the city jail and former Messerschmitt pilot, who evidently accompanied them by rail to Vienna. He is a tall, slender, with a hooked nose. He may be pro-Nazi but is friendly and brought back a note from Capt. John Dickey that said that he was living in Germany in a cell and that conditions were bad.

One of the Hungarian nurses, "Eva" (last name unknown) who was in love with one of the flyers, followed his shipment to Vienna where she learned that the flyers were imprisoned. She was not permitted to see them. "Eva" is a tall, slender, pretty, dark brown hair, brown eyes, cleft chin, aged about 21. She is known by Dr. Sukavarthy, who was in charge of the prisoners at the hospital.

On February 6, the day Buda was taken, the source requested the sergeant in charge of hospital records to make a complete list of Allied flyer patients for the 15th Air Force Battle Casualty section, which he promised to do within two days. However, source was wounded by machine gun fire the next day while going to get the list and was evacuated by the Russians. He was not permitted to return for the list.

APPENDIX F

S/Sgt. John Nagle, unharmed, and 2nd Lt. Elder A. Erfeldt, 2nd Lt.

Barrowcliff, 2nd Lt. Matthew Hendricks, 2nd Lt. Marshall Brown, and T/Sgt. John L. Lenburg, all wounded, were sent to Germany. These are members of source's crew.

A/C [aircraft] piloted by Capt. White was in formation with source's aircraft. Source saw this A/C blow up in mid-air with a very violent explosion, and believes all men were killed. No man from this crew came through the hospital.

2nd Lts. Scanlon, Becker and Friedman from one crew were taken prisoner. They told source that 2nd Lt. Evans crashed with A/C. This A/C was from source's squadron.

Daniel Whistler, from source's group (knocked down in the same raid) was shipped to Germany 17 November.

Pilot 2nd Lt. Richard L. Carroll and Thomas O'Connor (same crew); 2nd Lt. Hugh McGee (pilot), Jacques Le Poutre and Maurice Plante (same crew; 2nd Lt. Rodger Bullard and Gene Dosnier, and the navigator (same crew); pilot Lambert and Gunner Charles Robinson (same crew); 2nd Lt. Clyde L. Jones, Jr., 0-760324 (P-38 pilot); Captain John R. Dickey, 0-726341, Major Neil Lamont, 0-399730 (P-51 pilot); 2nd Lt. Fredris Rosemore (1523 First Ave. Nebraska City, Neb.); Sgt. Victor Lemle (828 Mason St., Toledo, Ohio), 2nd Lt. William Norhaus, Jr. (15 West 75th St., New York), 1st Lt. Henry Grove (2461 Amsterdam Ave., New York City); 1st Lt. Gary Johnson, Sgt. Robert Previto (Mobile, Ala.), 1st Lt. Harold Tomlinson (San Francisco, Calif.), 2nd Lt. Edward Wagner (Portland, Oregon); Sgt. Tony De Luca (Jersey City, N.J.); Sgt Samuel Nuccio, Sgt. Joseph Michaud (Springfield, Mass.), Sgt. James Lynch (North Bergen, N.J.), 2nd Lt. William Hiatt, Sgt. William Lamb (Tennessee or Kentucky), 1st Lt. Samuel Winfred (Florida), 1st Lt. Claude Rhodes, 1st Lt. Willard Graham (Greenwich, Conn.) died of neglect in hospital, Sgt. Paul Bertram (Minneapolis, Minn.), all passed through the hospital and were sent to Germany on or before 17 November 1944.

Other records of men are still in the military Hospital No. 11 at Budapest. Source had a man copying them but was shipped out too soon. <u>British Empire Personnel</u> encountered in the Royal Hungarian Hospital No. 11 by the source:

Serfontein, Capt. Francis	SAAF
Hamilton, Sgt. "Scotty"	RAF
Curtis, Sgt. Rupert	RAF
Patriarca, Sgt. Ron	RAF
Ludbom, F/O Hubert	RCAF
Thomas, Lt. Mike	Canadian Army

Thomas was known to the hospital as an American, though his English was very poor. He admitted to source his mission and circumstances of capture. From the time of their first meeting around 10 December 1944, until they were evacuated by the Russians 7 February 1945, they were in frequent communication. Source met Lt. Thomas again in Kecskemét, when they were billeted together until his departure for Italy 9 March 1945. Thomas was evacuated to Italy from Debrecen 17 March 1945.

End of Official Record

(This is a copy of a news article that was written by Paul Giguere that appeared in the newspaper the "Boston Globe" on May 20, 1945.)

Red 'Major' Home as a Plain Yank Sergeant

Maj. Leonard Bernhardt, Russian Army, who is also Staff Sgt Bernhardt of the United States Army, is back at his home on London Street, East, Boston.

Some Army buddies have taken to calling him "Major" but he still is a "sarge," although holding the rank of major in the Red Army.

His story begins on June 30, 1944, when the B-24 bomber of which he was a turret gunner exploded into flames under assault of German planes near Lake Balaton, Hungary. As the plane careened earthward, Sgt Bernhardt fought his way through the flames to the rear of the plane to help get out the tail gunner. The tail gunner was dead. By this time, the sergeant was badly burned, particularly about the face, and he had been hit by a bullet in his right leg.

Half unconscious, he bailed out and managed to get to the ground despite the attempt of a German pilot, zooming at the parachute, to try and cut the chutes shrouds with his plane's wings. Peasants wielding pitchforks and scythes swarmed around him and marched him away. When they noticed that his right leg was bloody, they hit the leg, and when he fell down, they kicked him to make him rise.

He was taken to the Royal Hungarian Honvedspital No.11 in Budapest, where he was kept on a stretcher. Doctors expected that he would die. On the fifth day, when it seemed he would recover, he was put in a bed. For many nights he was unable to sleep because his eyelids were gone.

At night a badly wounded stretcher case made a lasting imprint on Sgt. Bernhardt and gave him more courage than anything he has ever experienced in the war. This American soldier nearby never spoke during the day; only in his sleep, when he would say: "It doesn't hurt 'Mom'. It really looks worse than it is."

Russia's Red Army on the offensive in Belorussia in the summer of 1944. *Courtesy: WWIIToday.com*

The Red Army was approaching the city November 20 and the Germans evacuated the seventy-odd American patient, except for Sgt. Bernhardt. A member of the hospital staff, who belonged to the Allied underground, persuaded a German officer that Sgt. Bernhardt would lose his sight unless he remained at the hospital for an operation.

After he had recovered enough to walk around, the American airman became an active member of the underground movement eluding guards to operate a radio and to flash the location of ammunition dumps to the besieging Russians. Throughout this period, he subsisted on a bare menu, hardly enough to sustain him. Once in a bomb shelter (where he was wounded a second time by a mortar shell), somebody caught a dog and they feasted.

On February 6th, the Russians arrived at the hospital and he was surrounded by the Red Army men shouting, "Amerikanski". He was made a major in the Red Army for his underground work. He enjoyed a hearty meal. While he says the American Red Cross did everything possible to aid the lot prisoners, he believes the International Red Cross leader in the area was "corrupt".

The Russians were trying to storm the block next to the hospital. Sgt. Bernhardt picked up a sub-machine gun and joined in the assault. He lived and fought with Russian soldiers for the next two months, until a German machine gun bullet creased his scalp. In this period, his compatriots asked innumerable questions about America and pointed out name "Willys" cars and Boston (bombers), Russian doctors gave him the finest treatment before he left in an American plane which made a forced landing in the area, he was given the Soviet decoration "Order of the Guarde" The war is not over for him yet. He will have to undergo treatment for the next two years.

June 30, 1944: A local boy, Istvan Eros, surveys the wreckage of Captain White's plane shot down in action over Kisharsagy, Hungary. *Courtesy: Nándor Mohos*

WHITE'S CREW

Capt. John H. White, Jr. (P)
2nd Lt. Francis J. Ghiselli, Jr. (Bomb)
2nd Lt. Davis
2nd Lt. Northmore W. Hamill, Jr.*
S/Sgt. Wesley T. Cockroft (E)
Sgt. Anselm J. Cattoor (Ass/E)
S/Sgt. Lacy D. Powell (RO)
S/Sgt. John H. Blake (Ass\RO)*
S/Sgt. Donald C. Stevens (G)
S/Sgt. Richard J. Cole (G)
James F. Coble (photographer)
* killed in action

Captain White was flying the lead position in the high box on the June 30, 1944 Blechhammer mission to target one of two synthetic rubber and oil plants approximately two miles apart from each other in Blechhammer, German. This is a statement given by himself, Lacy Powell and Don Stevens:

Captain White received the return order, but poor visibility forced him to fly on. The three planes in our box re-entered the overcast, but as we emerged White realized his #3 engine was running rough, and the crew reported they had taken several rounds in the tail section from the attacking fighters. An attempt was made to join the group turning towards home, but because of loss of power, we could not keep up. We encountered light anti-aircraft fire, followed by a direct attack by several Me-210s.

Meantime, in the left waist position, radio operator/gunner Lacy Powell saw one fighter had peeled to the left and below the wing, and he fired one burst. Tracers indicated he had missed so he corrected and fired again. Tracers were spotted entering the enemy fighter, and photographer Coble shouted, "You got him!" Powell then assumed everything was all right and that they were heading back to base until a crew member reported an enemy fighter at 6 o'clock. The fighter circled out of range; then leveled off and "poured on the coal." When he got within firing range, he opened up and continued to fire until he hit the tail gunner's position, ignited bomb-bay gas lines, and caused an inferno.

Tail gunner Don Stevens had seen the single fighter coming in at them and kept firing, until the next thing he remembered was that he was lying in the fuselage looking up at the turret. Afraid of being trapped in the turret he had not closed the turret doors. That meant that, when the turret was turned to the extreme right or left position, his back was exposed to the wide-open spaces, but he accepted that as the lesser of the two evils.

The crew had reported three enemy fighters down or damaged, but the tail and belly turrets were no longer operative. In the next attack, there were several hits on the flight deck, bomb bay and waist, with fire breaking out. Then #3 engine quit, and #1 received a direct hit and exploded. Due to the extent of the fires, White gave the order to bail out and he set the plane on autopilot. Davis and Ghiselli were already on the flight deck. White made his way to bail out and found Cockcroft looking for his parachute, which the co-pilot had mistakenly taken. As Cockroft found a chute, the plane began to nose down. White went back to the controls and attempted to pull the nose up.

Top: A boy scout, with an American helmet in hand, stands in front of the Me-210 fighter downed by White's crew over Kisharsagy, Hungary. *Bottom:* Istvan Eros and a friend pose with one of the 20-millimeter cannons from the wrecked Me-210 plane used in battle. *Courtesy: Nándor Mohos*

Among the wreckage of the German Me-210 the plane's engine destroyed in combat. *Courtesy: Nándor Mohos*

Returning to the flight deck, White pushed Cockroft through the escape hatch just as an explosion occurred. White was thrown to the flight deck from the blast, but later realized that he was falling amid debris. He reached for his ripcord, which was not where he had expected it to be—then he remembered he had on a flak vest, so he pulled the cord on the vest and then on the chute and began floating down. On reaching the ground, he saw that the legs on his flight suit had burned off and his escape pack with them.

Powell had managed to find a parachute and had dropped out of the rear hatch. He too had trouble finding his ripcord as he had attached the clips upside down. Fortunately, it worked anyway. A fighter circled him but did not fire upon him.

When Don Stevens regained consciousness, the plane was filled with smoke, the escape door was open and everyone had bailed out so he put on his parachute and got out just before the plane exploded. On landing, he became aware of the shrapnel wound to the top of his right foot and a badly bruised right calf muscle.

Blake was killed in the plane, and Hamill was killed by peasants after he landed on the ground. Two of this crew were killed and nine became POWs.

Memorial honoring Lt. Evans and his crew killed in action on June 30, 1944.
Courtesy: John L. Lenburg Collection

EVANS'S CREW

2nd Lt. Robert G. Evans (P) *
2nd Lt. Stephen E. Mills (CP)
2nd Lt. Ralph C. Berger (N)
2nd Lt. Jerry Conlon (B)
S/Sgt. Cecil R. Boles (FE) *
Sgt. Walter W. Rowe (Ass\FE) *
T/Sgt. Guy Marsh Jr (RO) *
Sgt. Charles R. Becker
Sgt. Horace C. Barksdale (G) *
Sgt. Robert Friedman
* killed in action

Lieutenant Evans was flying in number three spot in our box formation on June 30, 1944 and his plane was one of the four shot down. The following was written by bombardier Jerry Conlon on Lieutenant Evans crew:

I was on the flight deck, sitting on the jump seat behind the pilot when there was the sound of explosions. I looked out the little window on the left side and could see German fighters, which I thought were Me-210s. Our left wing was badly damaged. Evans suddenly struck me across the chest with back of his hand yelling, "Get out of here!"

When I got into the bomb bay, Ralph Berger was attempting to crank open the bomb-bay doors without success. I told Jerry to get out of the plane and I started to the nose. Berger and Bob Friedman jumped out of the nose wheel hatch. I started back through the bomb bay, with Steve Mills behind me. The plane was hit by machine gun or cannon fire but was flying level. The rear deck was a complete mess. Waist gunners Boles and Marsh were sprawled out with terrible wounds. Horace Barksdale was lying backwards out of the rear turret with his legs still in the turret. Charley Becker was out of the ball turret, leaning over the waist gunners. I went to the other gunners, examined them, and found no signs of life from any of the three.

Second Lt. Robert G. Evans and his crew. *Courtesy: Jerry Conlon Collection*

Afterwards I jumped out from the camera hatch, from which Mills had already jumped. I'm not sure how Evans and Rowe died. They mortally either wounded before the crash or died upon impact. Five members of the crew were killed and five became POWs. The plane crashed on the property of Bodor Aladár and Tóth Mihály in Szigliget, a village in Veszprém county, Hungary.

"On Final" was written by Stephen E. Mills, co-pilot on this crew. As a civilian after the war, Steve flew small planes. While dying of cancer in 1977, Steve wrote this. It is read at the close of each of the 460th's reunions. I felt that it was appropriate to include it in this book.

ON FINAL

Whatever an aircraft is or has been—a vehicle of pleasure to fly among the birds, an instrument of war bursting with gun emplacement and bombs, or a record setter of speed, altitude and acrobatics—whatever its destiny, unfortunately, it is still a man-made machine that must return to earth and be tethered.

There is no other pilot maneuver that demands the full test of skill, depth perception, judgment, experience, alertness, than the final approach to landing. Anyone can fly a plane once it is airborne. They fly themselves, and with some skill and experience, even a novice can get a ship airborne.

But to land a plane, particularly if it is damaged, or at a strange field with limited facilities, short runway, power lines, rough runway or weather, bad winds—this is where the pro takes over to execute a smooth, safe landing.

On final, like life, is the last act before termination of a flight. I know of no other joy I have had than the thousands of final approaches I have made, each time trying to make it better than the last: Sudden crosswind—correction; a little too high—correction; too fast—correction. But the pure joy of command and flying of that aircraft on final is a pilot's report card.

I am "on final approach" of my life. I have few regrets. I have had many thrills and experiences, and I am pleased to hear the tower when they say: "Mills, you are #1 cleared to land."

On Palm Sunday, April 3, 1977, in the city of Wailuku on the island of Maui, Hawaii, Steve Mills made a perfect landing after cancer claimed his life.

CHAMPLIN'S CREW

2nd Lt. Nelson H. Champlin (P) *
2nd Lt. Edwin W. Baylor (CP)*
2nd Lt. Morton H. Osbom (N)
2nd Lt. Richard H. Whitaker (B)*
S/Sgt. Steve A. Marushok (E)*
S/Sgt. Arthur J. Popa (RO)*
Sgt. Eugene T. Potts Jr.
S/Sgt. Daniel E. Whistler (A)
S/Sgt. Howard E. Sexton*
S/Sgt. Howard L. Cales*
* killed in action

Lieutenant Champlin's plane of the 762nd Squadron had pulled into number five spot and was hit immediately by fighters. The number #3 engine caught fire and soon the bomb bay was in flames. Flames were leaping from the left gunner's window. The plane moved out

460th Bomb Group B-24 Liberators bomb enemy targets over Salzburg, Austria, in June 1944.

of formation to the right and disappeared between two cloud formations. At about 10,000 feet, it broke in two. Potts said that he was not able to tell anything about the people in the front of the plane after it broke in two. The tail of the ship glided until he was able to bail out. After determining Sexton was dead, Whistler bailed out. Cales jumped out without a parachute. His body was found later. This plane was one of the first planes to go down in our formation.

Seven were killed in action and three held as POWs. I met Dan Whistler at the 460th's Reunion in Albuquerque, New Mexico in September 1995. The last time that I had seen him was at the Royal Hungarian Hospital No. 11 in Tapolca, Hungary in June 30, 1944. His face had been burned very badly. He still carries the facial scars today.

Dan told me the attacking German fighters immediately hit their plane. The situation was so bad there was not time to get Cales out of the ball turret. With the plane now plunging to earth out of control, Dan bailed out. A Hungarian soldier, who saved him from the pitchforks of threatening farmers, took him prisoner. After spending months in Royal Hungarian Hospital No. 11, he was moved to Stalag Luft I.

Marvin Wycoff and Ray Swedzinski pictured at the 460th reunion in 1990 at Savannah, Georgia. *Courtesy: John L. Lenburg Collection*

SORGENFREI'S CREW

Kennon S. Sorgenfrei (P)
Raymond J. Swedzinski (Co-P)
Carl A. Pacharzina (N)
Joe Boczek (B)
Marvin Wycoff (R/O)
Paul Petersen
Stanley J. Radzierski
Edward Pyszczek
William R. Hensley
Theodore A. Turbak
Michael Bisek (Photographer - was only on the July 19, 1944 mission)

Lieutenant Sorgenfrei's plane and crew was a replacement crew in the 762nd Bomb Squadron. His plane was the only one able to escape the attack of German fighters on June 30, 1944 mission to Blechhammer. The following is a statement written by Marvin

Wycoff, a member of Lieutenant Sorgenfrei's crew who achieved the rank of Technical Sergeant and served as a radio operator and waist gunner. He writes about the effect this event of June 30, 1944 had on his life. At a 460th reunion in 1990 at Savannah, Georgia, I met Marvin for the first time. Since our first meeting in Savannah, Marvin and I have become good friends. Marvin wrote the following:

> *My wife and I attended the 1990 460th reunion in Savannah, Georgia, which was our first. Even though no other member of my flight crew was there, we found the reunion to be interesting and a very emotional experience. I was a radio operator/gunner on Ken Sorgenfrei's crew and we arrived in Spinazzola, sometime in April 1944. There was no longer any snow, but the mud was still alive and well as we made the 762nd Bomb Squadron our new home.*
>
> *We were a replacement crew assigned to the 460th at Spinazzola, a few weeks after the group arrived. Even though the squadron treated us well, we did not feel the same sense of belonging as the regular crews who had been together so long during training in the States. In retrospect, I believe this bonded our own crew members even more tightly as we found ourselves with different experiences and backgrounds than the original crews. Even our plane was different. I believe it to be the first silver plane on the field where all the other planes were painted olive drab. Our plane was used by other crews almost immediately and it did not survive long enough for us to use it in combat as it was lost about the time we were to fly our first mission. When it arrived, it was of considerable interest as it stood out like a silver star on the drab hardstand, glistening under the now bright Italian sun.*
>
> *We made a few practice flights, to confirm our stateside training, and began to fly combat missions with very little delay. We were on the mission to Vienna area on May 10, 1944 in*

which we, along with most of squadron, received a rather severe "roughing up" by flak. Our plane badly damaged, we limped back all the way from the target area alone, without seeing any enemy fighters, landing at Foggia with two crew members wounded. We used a flare as we approached, due to the command radio transmitter being out, to indicate our emergency situation and we would not have power to make a second attempt to land. Our radio receiver was working and we heard the tower instruct a Lancaster Bomber, in final approach, to go around allowing us to come straight in. Our landing, with a few problems, was successful and we just managed to get the plane off the runway where our two wounded, Joe Boczek and Bill Hensley, received medical attention. Joe and Bill were both lucky, however, as their wounds were not too serious and they missed only a few missions, getting back in the air again following a short rest and treatment.

We continued to fly missions at a rapid pace, skipping the rest period at Capri, in the hopes of finishing our 50 missions and return to the States. The road to this objective proved rough, with many obstacles along the way. One of these obstacles occurred on the trip to Blechhammer. If you were on this mission, you will remember the bad weather over Hungary and our formation entered the clouds placing us in extreme danger of mid-air collisions. I cannot remember where we were in the overall bomber formation, but Ken applied more power and climbed as rapidly as possible with our bomb load still aboard.

We cleared the clouds and found three other 460th bombers in a small formation. We attached to them as another B-24 approached, which I think was to our left-wing position; this made a box of five B-24s. There were no other 460th planes in sight at our higher altitude, probably near 20,000 feet. A long way off we could see a large formation, near our altitude and, even though I do not think our planes were talking by radio, our crew thought we were drifting closer to join this formation and perhaps drop our bombs on their target. As these planes got closer, however, they were not B-24s nor were they B-17s, they were a large formation of Me-210s, at least forty and they lost no time initiating a vicious attack. These fighters came in waves of four abreast from the rear with another wave following a short distance behind. They would open fire as soon as the first wave broke off their attack at a very close range. The plane beside us was hit hardest on these early waves and was soon burning and going down. This left us "Tail-end-Charlie" and now we were receiving the most attention from the Me-210s. We were taking a beating from their guns that were not machine guns but were 20-millimeter cannons. Other fighters, which I could not see from the waist position, came in from the front at the same time but the heaviest and steady attack was coming from the rear.

We had taken hits on the trailing edge of our wing, just behind the main landing gear, near engines one and two. There were large pieces of sheet metal in the flap areas just waving in the wind. We were soon running low on ammunition that these Me-210s never let our guns rest. Ken yelled something like, "Hold on! We're going down to the clouds," and about two seconds later, I was practically plastered to the roof of the plane. It took Ken and Ray Swedzinski to level our plane off in the clouds while still carrying a full bomb load. Here we played a deadly game of hide and seek for the next few minutes with the fighters going in out of the clouds. They would not come into the clouds with us and we found ourselves once again, in a badly damaged plane, trying to get back to Spinazzola, some two hours or more away.

We finally limped in, all the rest of the group had landed, and we were able to make a successful landing (even though the group called it a crash landing). During debriefing, we kept looking for the other planes in that high box. None returned. We were the only surviving plane of the five.

A short time later, July 19, 1944, we, too, were shot down over Munich. We tried to limp into Switzerland but found clouds over the mountains and since we still had enough altitude to clear, we delayed jumping until we could see the ground, however by this time we had over flown Switzerland and landed in the high Alps on the French side of the border. We were immediately picked up by French rural people and resistance fighters and spent about six weeks behind enemy lines.

The six weeks we were missing in France is completely another story in itself since we were with the French Resistance (FFI). We were chased by German patrols in the rough Alps as we helped care for a group of about twenty badly wounded Marquis members in a resistance hospital.

We finally got out when the 45th Division landed in southern France in August. The 45th drove rapidly up the Rhone valley, which allowed us to work our way back across the lines to American held territory. We hitched rides on trucks and a C-47, finally arriving back in Bari where they debriefed us, issued clean clothes, and sent back to Spinazzola for a few weeks and back to the United States in October.

This experience introduced us to "another war," even more vicious. There, we carried and helped care for patients with very severe wounds, amputations, etc. Sometimes we had nothing to eat. Since 1982, we have been in contact with this group and they now recognize us and have presented us with an: American Detachment" version of their French Marquis Flag. The French government considers us a part of the resistance movement and we are listed in a museum of the French Marines as contributing to their cause. The resistance group we were with was part of the French Marine Corps.

We have visited France twice, 1982 and 1985 and they have visited our group in 1983 and 1990.

Perhaps there are always parts of a struggle, such as we shared in 1944, which was so intense as to cause images to return throughout our lives and this MIA time in France was one of those times. The mission to Blechhammer, which was June 30, 1944, a date I did not know until our Savannah reunion, has always been close to the surface too and very frustrating to me as I knew nothing of the four crews who did not make it back. Following that summer of 1944, I talked to almost no one about these experiences. They were too painful and how could anybody possibly understand. Those feelings and frustrations are somewhat easier now since 1982 on the question of the MIAs, as we have re-established a contact with the people who shared our experiences.

Now in 1990, during the 460th reunion in Savannah, I accidentally mentioned the Blechhammer mission, and even though it seems impossible, I was talking to a survivor (Mike Brown) of that five-plane box.

Through him, I talked to people in three of the other four planes that were of the 760th Squadron. The three planes of the 760th had thirty-one crew members, including a photographer aboard. Twenty-one parachuted and were captured. The other ten did not survive the attack. These three planes were originally a part of the high box of the first attack unit and, though they knew one plane escaped, this was their first knowledge, too, of who we were.

We still do not know about one of the planes*, but I think it was probably the first one shot down which was likely on our left wing. If so, we saw it burning and I do not remember any parachutes but since we were busy with fighters and since there were clouds below us, maybe some could have escaped. Maybe next reunion will bring further information. The knowledge I *now have about the ten crew members from the 760th who gave their lives saddens me and the memories of that day will always be with me but there's now twenty-one survivors which I did not know of before and for that knowledge I'm grateful.*

*EDITOR'S NOTE: Wycoff is referring to Lieutenant Champlin's plane of the 762nd Bomb Squadron.

==

NOTE: Nineteen days later after their June 30th escape, Lieutenant Sorgenfrei's plane was shot down on a mission to Munich. I have reprinted their story that appeared in the *Readers Digest* in September 1984 under the title of "Saga of the Sorgenfrei Crew" I believe that you will find this story of what happened to this crew most interesting.

==

SAGA OF THE SORGENFREI CREW

By Andrew Jones

For the pilot Kennon Sorgenfrei and his bomber crew, July 19, 1944, marked a special occasion: the next-to-last combat mission of their tour in Europe. Soon they would return to their base in Spinazzola, Italy, and from there head across the Atlantic for home.

Their target - the railroad yards of Munich - was one of the hairiest on the continent; black anti-aircraft blossoms were everywhere around them. But they had been over it, and others as bad, many times. Now, with the tiny mesh of railroad tracks and freight cars in his sights, bombardier Joe Boczek tripped the bomb release and felt the B-24 jump as its load of 1000-pounders dropped free. Suddenly the plane jumped again, this time with a heart-stopping crash that told them they had been hit by flak.

"Joe," Sorgenfrei called over the intercom, "go aft and see what the damage is." Boczek scrambled back into the main compartment. There were no causalities, but two of the four engines were lost. Crippled, their bomber could never make it back to Italy.

Sorgenfrei told the navigator Carl Pacharzina to plot a course to neutral Switzerland. To lighten the ship, the crewmen jettisoned ammunition, flak suits, extra clothing. Meanwhile, their fighter escorts, low on fuel, rocked their wings, signaled good luck and broke off, heading for home. The stricken bomber was on its own.

The crew had absolute confidence in pilot Sorgenfrei and co-pilot Ray Swedzinski. Since their first days of training together in Casper,

Wyoming, they had been known as "the Sorgenfrei crew." With their tongue-twister names—Polish, Swedish, Jewish, Norwegian—they were proud of their polyglot all-Americanism. They were a well-disciplined outfit and knew from perilous experience three times their aircraft had been so badly damaged it had to be scrapped—that they could make it as a group. And so, peering down at the clouds that seemed to be creeping closer every minute, they set their jaws for whatever was coming next.

Suddenly one of the two remaining engines sputtered and quit. Sorgenfrei ordered everyone to bail out. Only seconds after they all jumped clear, the plane crashed and exploded with an earthshaking roar. The time margin was so close that the last man out landed 200 yards from the wreckage.

Even as they were scrambling out of their harnesses, local farmers surrounded them. Turret gunner Paul Petersen, assuming they were in the Swiss Alps, shouted at one, "Swiss?"

"None!" the man answered. "Francais!"

With their navigational equipment malfunctioning and the clouds obscuring their view, they had overflown Switzerland and come down in German-occupied Vichy France.

Quickly, the locals herded the eleven crewmen into a wood at the top of a hill. They had landed between two German garrisons and practically on top of a road that was regularly patrolled. Within 45 minutes, as they watched from the hilltop, the smoking ruins of their plane were surrounded by gendarmes. German troops arrived later and were so frustrated at the American's disappearance that they emptied their machine guns into the wreckage.

In one way, however, the crew was lucky. They had come down near Prunières, a village some 50 miles southeast of Grenoble, in an area where large groups of patriots had organized into a resistance network, known as Le Marquis, that was giving the Germans a difficult time. The maquisards [guerilla fighters in the French underground] promised to do everything possible to get the Americans to Switzerland.

None of the crew spoke French, and the peasants knew little English, but through gesture and repetition, they hammered out a plan. The Americans spent two nights on the forested hilltop. On the third day, they were packed into a truck and driven north toward Switzerland.

Their first meal was near the village of Ancelle, at the home of a Marquis leader named Joseph Brochier. The next morning, maquisard mountaineer Laurent Artru and a professional guide set out with the crewmen on a three-day trek across a glacier to La Bérarde, a relatively safe village in the masif. There, the Americans were turned over to another resistance group, the Maquis de l'Oisans, commanded by the thunder-voiced Le Capitaine André Lanvin of the French colonial forces. Lanvin had been recalled from French-Indochina at the beginning of the war and had brought a full company of Vietnamese soldiers with him. When the Vichy government surrendered to the Germans, he and his 250 men had taken to the hills. Now they defended this mountain stronghold against 10,000 German troops. With characteristic disdain for the Germans, maquisards and villagers publicly welcomed the flyers with champagne and flowers.

The next leg of the American's journey took them to the small abandoned ski resort of l'Alpe d'Huez, where the Maquis de l'Oisans had set up a hideaway field hospital in the former ski lodge. There the Americans helped twenty wounded underground fighters. They also met Noel Monod, a wealthy, patrician Parisian, who took charge of the flyers then on.

By now, ten days after shoot-down, the Germans were torturing and shooting any prisoners suspected of withholding information about the flyers whereabouts. Meanwhile, Monod and Lanvin were receiving intelligence about German troop movements from a number of sources, including 20-year-old Pierre Montaz, who was working at l'Alpe d'Huez. Many suspected that the Germans were planning a massacre of the Maquis de l'Oisans as if they had just inflicted on nearby Vercors.

For the Sorgenfrei crew, getting to Switzerland now became secondary to their work in occupied France. They feared that because of their association with the marquisards, if captured, they would be considered resistance fighters rather than prisoners of war and get a firing squad instead of a POW camp. They were not joining in Lanvin's raids, blowing up bridges, trains and highways, but their help around the hospital was a valued contribution to the resistance cause.

They remained 12 days. Then one morning came the dreaded but not unexpected news: the Germans were on their way. Immediate evacuation of the hospital was ordered—straight up into the mountains. Hastily, the group stripped the building of whatever essentials could be carried by three mule carts.

For three days and nights, they climbed ever higher, dodging behind rocks as Stuka dive-bombers roared overhead through the mountain passes. On the third day, they came upon terrain so rugged they had to abandon the carts and animals. The crewmen took turns carrying two leg-amputees on stretchers.

That afternoon, the lower slope swarmed with gray-green uniforms of a German patrol, and automatic-weapons fire zinged around the fugitives' heads. Clearly, there could be no escape with the stretcher cases. In the next heartbreaking minutes, the stretchers laid side by side in a little declivity, and a cairn of rocks piled around them in hope that the Germans would not find them. A male nurse from the hospital and the fiancée of the amputees stayed with them. By now, the maquisards were returning the fusillade and a full-scale firefight was raging.

That night, after the skirmish subsided, Monod and a crew gunner Paul Petersen returned to the cairn and prayed with the two amputees. Monod then offered them his revolver. "No," one man said. "Christ does not believe in suicide." Monod turned and started back down the mountain.

The most demanding leg of the journey lay ahead. They were often 10,000 feet above sea level, often without food or shelter. Even in August, temperatures after sundown dropped to freezing. Dehydrated, weakened by hunger, they pressed on.

Then suddenly the fortunes of the war changed—and so did their own. Monod learned that the Allies had landed on the coast of Provence and had broken the German stranglehold on south-central France. Grenoble came under Allied attack in mid-August. From their perch up among the glaciers, the group could hear artillery fire.

Finally, thirty-seven days after their crash, the Americans walked down the mountain and into liberated Grenoble. There was citywide celebration and a hero's reception. To add to their joy, the crew learned that the two amputees were safe. The German patrol had passed within fifty feet without seeing them, and they had been rescued by their Marquis comrades.

Thus, the war ended for the Sorgenfrei crew. Several days later, they returned to their home base in Italy.

In the following years, the crewmen went their various ways: Ken Sorgenfrei became a commercial airline pilot, Ray Swedzinski went back to his family's Minnesota farm, Paul Petersen became a Lutheran minister. Some of them kept in Christmas card touch, but they made no effort to get together and talk over old times.

Ten, in February 1982, Sorgenfrei received an invitation from France; a ceremony held at a monument built in the mountains to honor the Maquis de l'Oisans, and he and his crewmen were invited to attend. Three of them made it: Sorgenfrei, navigator Carl Pacharzina and radio operator Marvin Wycoff. For their part in carrying French wounded across the Alps, each received a medal commemorating the liberation of Grenoble, and all the crewmen became honorary members of the Marquis de l'Oisans.

The trip to France broke the 38-year separation of the Sorgenfrei crew. On July 7, 1983, eight of the eleven original crewmen and a small group of their French rescuers gathered at the Kirkwood Motor Inn in Bismark, N.D., Marvin Wycoff's hometown. Noel Monod, retired treasurer of the United Nations, arrived from New York City. From France came Laurent Artru, head of the Chamber of Commerce in Lyon, and Pierre Montaz, a ski manufacturer in Grenoble.

In the next four days, the American showed their guests around North Dakota. They had breakfast with the state's governor and lieutenant governor and received a congratulatory telegram from Secretary of Defense Caspar Weinberger on behalf of President Reagan. Pierre Montaz showed slides, retracing the crew's route through the mountains and showing many of the people who had offered assistance nearly four decades before. Noel Monod shared snapshots he had taken during the trek itself, using film "borrowed" from the Germans.

There were light moments: the Americans laughed to hear that, immediately upon liberation, the women of the area where they had bailed out appeared in the streets wearing beautiful blouses of the white parachute silk. And there were quiet moments: at Sunday morning mass all thoughts were on those who gave their lives for their homelands and helped ensure that a handful of Yankee airmen would make it back safely through the occupied country to freedom in their own.

However, the story of the Sorgenfrei crew did not end there on the Dakota plains. On the group's last day together, plans were made for a trip to France in the summer of 1985 to retrace the escape route. In a speech of temporary farewell, Montaz lent poetic perspective to their long-ago wartime saga. "The ocean is between us," he said, "but history unites us."

Footnote: Twenty-eight planes were dispatched on the Munich mission. Twenty planes returned to the base. Fifteen of these were damaged by flak. Fifteen men were killed and forty-two became POWs.

With Jack Nagle seated the front porch of his father's house in Houston.
Courtesy: John L. Lenburg Collection

JACK NAGLE

This is a statement given by Jack Nagle, our nose turret gunner:

The Blechhammer mission was to be a different mission. The target would be farther than our normal range. As we flew through the clouds over Hungary, they recalled our mission due to bad weather. As we broke out of the clouds, a group of German fighter planes was waiting for us. They knocked down four of the five planes in our box. Our plane was on fire, the intercom was gone, and the navigator was banging on my nose turret door, giving me the sign to get out. I opened the door and started to back out of my turret. Suddenly I spotted a German fighter just off our left-wing tip. I turned the turret and blew him away. The navigator then opened the nose wheel door for us—to squeeze our way through

Me with Jack (second from the right) with our fellow crew members (from left to right): Ralph Wheeler, Leonard Bernhardt, Martin Troy, and Rube Waits, Jr. *Courtesy: John L. Lenburg Collection*

the tight opening—and parachute from the plane. When I parachuted out, we were over Lake Balaton. I landed 200 feet from the lake's edge, a farmer with an old musket immediately fired at me. Fortunately, he only had one bullet and he missed.

Other farmers in the area, armed with pitchforks and scythes, held me there until a German soldier picked me up. Shortly thereafter, the seven members of my crew and I reunited: the ball turret gunner the waist gunner and the tail gunner had been killed.

Next, I moved to a civilian prison in Budapest. There it was one man per cell. We got a cup of hot water and a small bread roll in the morning. I thought it was breakfast, but it turned out that was my ration for the day.

The Germans took my wedding ring, pocket comb and flight suit. After several days, the interrogations began. The interrogator was an English-speaking German pilot, who—as fate would have it—had attended the same

New Jersey high school as one of my crew members. Unfortunately, this coincidence brought me no leniency.

The only information I gave the Germans was his name, rank and serial number, which earned me 21 days in solitary confinement. When hunger and bedbugs finally got the best of me, I told the interrogator the names of my crew members. The German then showed me a file. They knew every place I had trained in the U.S. as well as all the crew members' names. He told me he had to confirm that I was not a spy. Spies were shot.

Jack in his co-pilot seat on "Miss Fortune." *Courtesy: John L. Lenburg Collection*

Other POWs and I loaded ourselves into cattle cars for the long trip to Germany. There was no food or water, no sanitary facilities, nor was here enough room for all of us to lie down, so we took turns. I also remember a terrifying night in Vienna, when bombs hit railcars on both sides of us by the RAF.

When we arrived at the prison camp, Stalag Luft IV, we were unloaded, handcuffed and forced to run to the camp. I became Prisoner of War, #6420.

CAPT. LESLIE CAPLAN

Capt. Leslie Caplan was a flight surgeon in the 15th Air Force assigned to the 449th Bomb Group. Flight surgeons were required to fly combat missions, even though they did not play any role as a participating crew member during the mission but rode along only as an observer. This gave them an insight of the rigors and mental strain put on the men flying these missions. The plane he was on was shot down over Yugoslavia in October 1944.

Captain Caplan was captured and eventually shipped to Stalag Luft IV. The Germans sent him there because he was an American medical doctor even though the camp was for enlisted men. During the march he was the medical doctor in charge Section C (the section to which I was assigned).

On March 28, 1945 after fifty days of walking, Section C was split into two groups and loaded fifty men to a boxcar. The group with Captain Caplan was shipped to Stalag IIB near Fallingbostel and the other group with me was shipped to Stalag XIA near Altengrabo. I was able to obtain a copy of Captain Caplan's testimony to the War Crimes office in December 1947 about the conditions endured by the POWs at Stalag Luft IV and on the march. He certainly tells it like it was.

FOR THE WAR CRIMES OFFICE, CIVIL AFFAIRS DIVISION, DSS
UNITED STATES OF AMERICA

In the matter of mistreatment of Perpetuation of Testimony of American prisoners of war, Stalag Luft IV, Dr. Leslie Caplan (Formerly Major from November 1944 to May 1945. MC, ASN 0-413434)

Taken at:	Minnesota Military District, The Armory, 500 So. 6th St. Minneapolis, 15, Minn.
Date:	31 December 1947
In the Presence of:	Lt. Col. William C. Hoffman, AGD Executive Officer, Minnesota Military District, The Armory, 500 So. 6th St. Minneapolis, 15, Minn.
Questions by:	Lt. Col. William C. Hoffman, AGD

Q. State your name, permanent home address, and occupation.
A. Leslie Caplan, Dr., 1728 Second Ave. So., Minneapolis, Minnesota; Resident Fellow in Psychiatry, University of Minnesota & Veterans Hospital, Minneapolis, Minn.

Q. State the date and place of your birth and of what country you are a citizen.

A. 8 March 1908, Steubenville, Ohio; citizen of the USA.

Q. State briefly your medical education and experience.
A. Ohio State University, BA, 1933; MD, 1036; University of Michigan postgraduate work in Public Health; University of Minnesota Graduate School; one-year internship, Providence Hospital, Detroit, Michigan; 4 years general practice of medicine in Detroit, Michigan, 1937-1941; 4 years Flight Surgeon, U.S. Army, 1941-1945.

Q. What is your marital status?
A. I am married

Q. On what date did you return from overseas?
A. 29 June 1945

Q. Were you a prisoner of war?
A. Yes

Q. At what places were you held and state the approximate dates?
A. Dernisch, Jugo-Slavia 13 October 1944 to 20 October 1944; Zagreb, Jugo-Slavia 27 October 1944 to 1 November 1944; Dulag Luft, Frankfort, Germany, 15 November 1944 to 22 November 1944; Stalag Luft IV, 28 November to 6 February 1945; on forced march under jurisdiction of Stalag Luft IV February 1945 to 30 March 1945; Fallingbostel Stalag II B March 30, 1945 to April 6 1945; on forced March from 6 April 1945 to 2 May 1945.

Q. What unit were you with when captured?
A. 15th Air Force, 449 Bomb Group, 719th Squadron. I was Flight Surgeon for the 719th Squadron.

Q. State what you know concerning the mistreatment of American prisoners of war at Stalag Luft IV.
A. The camp was opened about April 1944 and was an Air Force Camp. It was located at Gross Tyschow about two miles from the Kiefheide railroad station. In the summer of 1944, the Russian offensive threatened Stalag Luft VI, so approximately 2000 Americans were placed on a ship for evacuation to Stalag Luft IV. Upon arrival at the railroad station, certain groups were forced

to run the two miles to Stalag Luft IV at the points of bayonets. Those who dropped behind were either bayoneted or were bitten on the legs by police dogs.

Q. Were the wounds serious enough to cause any deaths?
A. All were flesh wounds and no deaths were caused by the bayoneting.

Q. Did you see these men at the time of the bayoneting?
A. No. This happened prior to my arrival at Luft IV.

Q. Did you see any of the men who were bitten by dogs?
A. Yes, I personally saw healed wounds on the legs of a fellow named Smith or Jones (I am not certain of the name) who had been severely bitten. There were approximately fifty bites on each leg. It looked as though his legs had been hit with buckshot. This man remained an invalid confined to his bed all the time I was at Stalag Luft IV.

Q. Do you know how many men were injured, as a result of the bayonet runs?
A. I was told that twenty men had been hospitalized as a result. Many other bayoneted men were not hospitalized due to limited facilities.

Q. Who told you of these incidents?
A. Captain Wilbur E. McKee, 1462 So. Seventh St., Louisville, Ky., who was chief Camp Doctor. He should have some authentic records. Captain Henry J. Wysen, 346 E. Havenswood Ave., Youngstown, Ohio, also knew of the incidents. There were also two enlisted men who were elected by the soldiers as camp leaders and known officially as "American Man of Confidence," who could give an account of the camp bayoneting. The chief "American Man of Confidence" was camp leader and should have complete records of the incident. His name is Frank Paules, 101 Regent St., Wilkes-Barre, Pennsylvania. Francis Troy, Box 233, Edgerton, Wyoming, the other enlisted man, and "American Man of Confidence" should also verify incidents. Both of these enlisted men were also on the forced march when Stalag Luft IV was evacuated.

Q. Do you know if the Commandant was responsible for the bayoneting and dog bites?
A. I did not know the Commandant and do not know who was responsible. Captain Pickhardt, the officer in charge of the guards, is said to have incited the

guards by telling them that American airmen were gangsters who received a bonus for bombing German children and women. Most of the guards were older and fairly reasonable, but the other guards were pretty rough. "Big Stoop" was the most hated of the guards.

Q. For what reason was "Big Stoop" disliked?
A. He beat up on many of our men. He would cuff the men on the ears with an open hand sideways movement. This would cause pressure on the eardrums that sometimes punctured them.

Q. Could you give specific incidents of such mistreatment by "Big Stoop"?
A. Yes. I treated some of the men whose ear drums had been ruptured by the cuffing administered by "Big Stoop".

Q. Can you describe "Big Stoop"?
A. He was about six feet, six inches tall, weighed about 180 or 190 pounds, and was approximately fifty years old. His outstanding characteristic was his large hands, which seemed out of proportion to those of a normal person.

Q. When you arrived at Stalag IV, were you subjected to the bayonet runs?
A. No. We were marched from the station to Luft IV, but not on the run. Some of the men were tired and we did complain to "Big Stoop". He snarled at us, but personally went forward slowed the column down.

Q. Did you have duties assigned to you while a prisoner?
A. I was known as an Allied Medical Officer at Stalag Luft IV Camp Hospital and in charge of Section C while on the march.

Q. State what you know concerning the forced march from Stalag Luft IV?
A. In February 1945, the Russian Offensive threatened to engulf Stalag Luft IV. On 6 February 1945, about 6,000 prisoners were ordered to leave the camp on foot after only a few hours' notice. We left in three separate sections, A, C, and D. I marched with Section C, which had approximately 2,500 men. It was a march of hardship. For 53 days we marched long distances in bitter weather and on starvation rations. We lived in filth and slept in open fields or barns. Clothing, medical facilities and sanitary facilities were utterly inadequate. Hundreds of men suffered malnutrition,

exposure, trench foot, exhaustion, dysentery, tuberculosis, and diseases. No doubt, many men are still suffering today as a result of that ordeal.

Q. Who was in charge of this march?

A. The commandant of Stalag Luft IV was in charge of three sections. Hauptman (Captain) Weinert was in charge of Section C that I marched with. All the elements of Stalag Luft IV occupied a good bit of territory and there frequent overlapping of the various sections.

Q. How much distance was covered in this march?
A. While under the jurisdiction of Stalag Luft IV, we covered an estimated 555 kilometers (330 miles). I kept a record, which I still have of distances covered, rations issued, sick men abandoned, and other pertinent data. This record is far from complete especially about records of the sick, but the record of rations and distances covered is complete.

Q. How much food was issued to the men on this march?
A. According to my records, during the fifty-three days of the march, the Germans issued us rations that I have figured out to contain a total of 770 calories per day. The German ration was mostly in potatoes and contained very little protein, far from enough to maintain strength and health. However, in addition we were issued Red Cross food which, for the same fifty-three-day period, averaged 566 calories per day. This means at our caloric intake per day on the march amounted to 1336 calories. This was far less than the minimum required to maintain body weight, even without the physical strenuous activity we were compelled to undergo in the long marches. The area we marched through was rural and there were no food shortages there. We all felt that the German officers in our column could have obtained more supplies for us. They contended that the reason we received so little Red Cross supplies was because the Allied Air Force (of which we were "Gangster" members) had disrupted the German transportation that carried Red Cross supplies. This argument was disproved later when we continued our march under the jurisdiction of another prison camp; namely Stalag IIB. This was during the last month of the war when German transportation was at its worst. Even so, we received a good ration of potatoes almost daily and received issues of Red Cross, far more than were given under the jurisdiction of Stalag Luft IV.

Q. What sort of shelter was provided during the 53-day march?
A. Mostly, we slept in barns. We were usually herded into these barns so closely that it was impossible all men to find room to lie down. It was not unusual many men to stand all night or be compelled to sleep outside because there was no room inside. Usually there was some straw for some of us to lie on but many had to lie in barn filth or in dampness. Very frequently there were large parts of the barn (usually drier and with more straw) that were denied to us. There seemed to be no good reason why we should have to sleep in barnyard filth or stand in a crowded barn while other sections of the barn were not used.

The Germans sometimes gave no reason for this but at other times, it was made clear to us that if we slept in the clean straw its value to the animals would be less because we would make it dirty. At other times, barns were denied to us because the Germans stated having POWs in the barn might cause a fire that would endanger the livestock. It was obvious that the welfare of the German cattle was placed above our welfare. On 14, February 1945, Section "C" of Stalag Luft IV had marched approximately thirty-five kilometers (21.74 miles).

There were many stragglers and sick who could barely keep up. That night the entire column slept in a cleared area in the woods near Schweinemunde. It had rained a good bit of the day and the ground was soggy; but it froze before morning. We had no shelter whatsoever and were not allowed to forage for firewood. The ground we were to sleep on was littered with feces from prisoners who had previously stayed there with dysentery. There were many barns in the area, but no effort was made to accommodate us there. There were hundreds of sick men in the column this night. I slept with a POW who was suffering from pneumonia.

Q. What were the conditions on this march as regards drinking water?
A. Very poor. Our sources of water were unsanitary surface water and well water. Often questionable sanitary quality. At times so little water was issued to us that men drank whatever they could. While there was snow on the ground, it was common for men to eat snow whether it was dirty or not. At other times, some men drank from ditches that others had used as latrines. I personally protested this condition many times. The doctor from Stalag Luft IV (Capt. Sommers or Sonners) agreed that the lack of sanitary water was the principal factor responsible for the dysentery that plagued our men. It would have been a simple matter to issue large amounts of boiled water that would have been safe regardless of its source. At times, we were issue adequate amounts of boiled

water but at other times, not enough safe water was available. We often appealed to be allowed to collect firewood and boil water ourselves in many boilers that were standard equipment on almost every German farm. This appeal was granted irregularly. When it was granted, the men lined up in the cold for hours to await the tedious distribution. Another factor that forced an unnecessary hardship on us was the fact that when we first left Stalag Luft IV, the men were not permitted to take along drinking utensil. The first few issues of boiled water were therefore not widely distributed for there were no containers for the men to collect the water in.

As time went on, each man collected a tin can from the Red Cross food supplies and this filthy container was the sole means of collecting water or the soup that was sometimes issue to us.

Q: What medical facilities were available on the march from Stalag Luft IV?
A. They were pitiful. From the very start, large numbers of men began to fall behind. Blisters became infected and many men collapsed from hunger, fear, malnutrition, exhaustion, or disease. We organized groups of men to aid hundreds of stragglers. It was common for men to drag themselves along in spite of intense suffering. Many men marched along with large abscesses on their feet or frostbite of extremities. Many others marched with temperatures as high as 105 degrees Fahrenheit. I personally slept with men suffering from erysipelas, diphtheria, pneumonia, malaria, dysentery and other diseases.

The most common disease was dysentery for this was inevitable consequence of the filth we lived in and the unsanitary water we drank. This was so common and so severe that all ordinary rules of decency were meaningless. Hundreds of men on this march suffered so severely from dysentery that they lost control of their bowel movements because of severe cramps and soiled themselves. Wherever our column went, there was a trail of bloody movements and discarded underwear (which was sorely needed for warmth). At times, the Germans gave us a few small farm wagons to carry our sick. The most these wagons ever accommodated was thirty-five men, but we had hundreds of men on the verge of collapse. It was the practice to load the wagon. As a man would collapse, he would be put on the wagon and some sick man on the wagon would be taken off the wagon to make way for his exhausted comrade. When our column would near a permanent POW camp, we were never allowed to leave all of our sick. I do not know what happened to most of the sick that were left at various places along the march.

Q. What medical supplies were issued to you by the Germans on the march from Stalag Luft IV?

A. Very few. When we left the camp, we carried with us a small amount of medical supplies furnished us by the Red Cross. At times, the Germans gave us pittance of drugs. They claimed they had none to spare. At various times, I asked for rations of salt. Salt is essential for the maintenance of body strength and of fluids and minerals. This was particularly needed by our men because hundreds of them had lost tremendous amounts of body fluids and minerals as the result of dysentery. The only ration of salt that I have record of or can recall was one small bag of salt weighing less than a pound. This was for about 2500 men. I feel there is no excuse for this inadequate ration of salt.

Q. To your knowledge, did any sick man die, as a result of neglect by the Germans on the march from Stalag Luft IV?

A. Yes. The following men died as a result of neglect. All of these men have been declared dead by the Casualty Branch of the Adjutant General's Office:

NAME	ASN	GRADE
George W. Briggs	39193615	S/Sgt.
John C. Clark	33279680	S/Sgt.
Edward B. Coleman	12083472	S/Sgt.
George F. Glover	16066436	S/Sgt.
William Lloyd	18217669	S/Sgt.
Harold H. Mack	17128736	T/Sgt.
Robert M. Trapnell	13068648	S/Sgt.

It is likely that there were other deaths that I do not know about.

Q. Did all these deaths occur while the men were directly under control of Stalag Luft IV?

A. No. As I mentioned before, our sick men were left at various places and I never saw them again. Some of these men died after we were out of the jurisdiction of Stalag Luft IV.

Q. What were the circumstances that led to the deaths of these men?

A. At 0200 on 9 April 1945 at a barn in Wohlen, Germany, Sgt. George W. Briggs was suddenly overcome by violent shaking of the entire body and soon after went into a coma. This patient was sent to a German hospital. We were then under the jurisdiction of POW Camp Stalag IIB and they voluntarily sent this patient to a hospital. This is in marked contrast to the treatment received when we were under jurisdiction of Stalag Luft IV when every hospitalization was either refused or granted after a long series of waiting for guards, waiting for permission to see Capt. Weinert, and waiting his decision. In spite of the prompt hospitalization, this patient dies on 11 April 1945. No doubt, this death was largely caused by being weakened on the first part of the march while under the jurisdiction of Stalag Luft IV. On 9 March 1945, while on the march in Germany, Capt. Sommers who was the German doctor for Stalag Luft IV, personally notified me that John C. Clark had died the previous night of pneumonia.

He had not been hospitalized and had received very little medical care. I never saw this patient, but he was seen in a barn in the terminal stages of his illness by Capt. Pollack of the Royal Medical Corps who told me about it later on.

On 13 April 1945, while on the march in Germany, Edward B. Coleman collapsed from severe abdominal pain and weakness. I made a diagnosis of acute abdominal emergency superimposed on a previously weakened condition that was the result of malnutrition and dysentery. He was hospitalized but, according to the records of the Adjutant General, he died 15 April 1945.

On 14 April 1945, George F. Grover was seriously ill and an officer from Stalag IIB authorized him to be sent to a German hospital. He was suffering from intestinal obstruction, exhaustion, malnutrition and dysentery, mostly the result of mistreatment while under the jurisdiction of Stalag Luft IV.

About 8 March 1945 while on the march in Germany under the jurisdiction of Stalag Luft IV, I set up a resting place for the sick at a barn in Beckendorf. Harold W. Mack was carried into this barn suffering from dysentery, malnutrition, exhaustion, frostbite and impending gangrene of both feet. Permission to send him to a hospital was denied by Capt. Weinert. On March 9th, our column was ordered to march about six kilometers. Sgt. Mack, and many others, was too weak to march so he was placed on a wagon and taken along. He was so weak at the time that he had to be spoon-fed and had to be carried to the latrine. On March 10th, after another appeal to Capt. Weinert, Sgt Mack was sent to Beckendorf to await shipment to a German hospital, Sgt. Mack had both feet amputated. According to the records of the Adjutant General, Sgt. Mack died in Germany 2 April 1945. Sgt. King had all his toes amputated at the same German hospital, but he recovered.

On 24 February 1945, I was operating a barn hospital at Bradenfeld, Germany under the jurisdiction of Stalag Luft IV on the march. Sgt Trapnell was a patient at this hospital suffering from dysentery and exhaustion. In addition, he developed symptoms of acute appendicitis, which required surgery. Capt. Weinert authorized me to transport this patient by wagon to what he called a hospital at a nearby village of Bryge (or Brige). He must have known that a village of only a few people would not have a hospital. When I arrived at Bryge, I found that the so-called hospital was a barn with no medical facilities. Capt. Hay of the Royal Medical was in charge of the sick there. He agreed with me that Sgt Trapnell was seriously ill and that his acute appendicitis warranted immediate surgery. We had no anesthetics or other supplies, not even a knife. We were both covered with filth. Capt. Hay hoped that he would be allowed to send Sgt. Trapnell to a German hospital the next day. I do not know how long it took to send Sgt. Trapnell to a hospital for I had to rejoin my column at once. The records of the Adjutant General state that Sgt. Trapnell died on 5 March 1945.

Q: Do you know of any other men who were seriously harmed by this march from Stalag Luft IV?
A. Yes. There must be hundreds of men still suffering as a result of the rigors of that march. I personally tended to hundreds of such men on the march. I still hear many of them and there are numerous complaints about their health. I will cite a few instances. I know of three men who suffered pulmonary tuberculosis after the march. No doubt, there were many others that I never knew about. I was evacuated from the ETO on the hospital ship "ACADIA" and on that one boat there were over 20 men from Stalag Luft IV. One of these was S/Sgt. Norman C. Edwards, ASN 33558570 of Baltimore, Maryland. He was one of the men left behind during the march from Stalag Luft IV. Sometime in March or April 1945 he had both legs amputated because of gangrene secondary to frostbite. He told me that S/Sgt Vincent Soddaro, ASN 32804649 of Brooklyn, New York had also both legs amputated because gangrene and frostbite. Sgt. Edwards and Sgt. Soddaro had been in the same German hospital.

Q. What other mistreatment did you suffer on the march from Stalag Luft IV?
A. There were beatings by the guards at times, but it was a minor problem.

At 1500 hours on 28 March 1945, a large number of our men were loaded on freight cars at Ebbsdorf, Germany. We were forced in at the rate of sixty men or more to a car. This was so crowded that there was not enough room for all men to sit at the same time. We remained in these jammed boxcars until 0030 hours 30 March 1945 when our train left Ebbsdorf. During this 33-hour period, few men were allowed out of the cars for the cars were sealed shut most of the time. The suffering this caused was unnecessary for there was a pump with a good supply of water in the railroad yards a short distance from the train. At one time, I was allowed to fetch some water for a few of our men who was suffering from dysentery. Many men had dysentery at the time and the hardship of being confined to the freight cars was aggravated by the filth and stench resulting from men who had to urinate and defecate inside the cars. We did not get off these freight cars until we reached Fallingbostel around noon of 30 March 1945 and then marched to Stalag IIB. The freight cars we were transported in had no markings on them to indicate that they were occupied by helpless prisoners of war. There was considerable aerial activity in the area at the time and there was a good chance of being strafed.

Q. Was the suffering that resulted from the evacuation march from Stalag Luft IV avoidable?
A. Certainly, a large part of the suffering was avoidable. As I mentioned before, marched through rural Germany and there was no lack of food there. There was always firewood available that could have been used to boil water and thus give us a supply of safe drinking water. There were many horses and wagons available that could have been used to transport our sick men. There were many men in our column who were exhausted and who could have been left for a rest at prison camps that we passed on the march.

On 30 March, we left the jurisdiction of Stalag Luft IV when we arrived at Stalag IIB. On 6 April 1945, we again went on a forced march under the jurisdiction of Stalag IIB. Our first march had been in a general westerly direction for the Germans were then running from the Russians. The second march was in a general easterly direction for the Germans were then running from the American and British forces. Because of this, during the march under the jurisdiction of Stalag IIB we doubled back and covered a good bit of the same territory we had just come over a month before. We doubled back for over 200 kilometers and it took 26 days before British forces liberated us. During those 26 days, we were accorded much better treatment.

We received a ration of potatoes daily besides other food, including horsemeat. We always had barns to sleep in although the weather was much milder than when we had previously covered this same territory. During those 26 days, we received about 1235 calories daily from the Germans and an additional 1500 calories daily from Red Cross for a total caloric intake of about 2735 calories a day. This was far more than we had in the same area from Stalag Luft IV. I believe that if the officers of Stalag Luft IV had made an effort they too could have secured us as much rations and shelter.

Q. To what officers from Stalag Luft IV did you complain?
A. I only saw the commandant of Stalag Luft IV once on the entire march and was not allowed to talk to him then. Mostly I complained to Capt. Weinert who was in charge of "C" column that I was with most of the time.

Q. Can you describe Capt. Weinert?
A. He was a little taller than average and well built. He was in his forties but looked much younger until he took off and exposed his baldhead. He was an Air Corps officer and was said to have been a prisoner of the Allies in North Africa and later repatriated for a physical disability. I never saw any certain evidence of such a disability. He rarely marched but rode in his own wagon.

Some of the men said he had an arm injury but I never saw any definite evidence of this. Maybe this was because I only saw him on rather formal military occasions when he would stand or sit in a rigid manner almost as if he were at attention. I never saw him for long periods of time. He spoke excellent English, but it was a favorite trick of his to act as if he did not understand English. Usually he spoke to me through an interpreter, but several times, we spoke in English.

Q. Are there any other incidents that should be reported.
A. There is one incident I would like to report. On 16 February 1945, we were on the road west of the Oder River in the general area of Schweinemunde. I was marching with a party of several hundred of our stragglers who were tagging along behind our main column. We met a small group of other prisoners on the road.

I was allowed to talk to these men briefly and obtained the following information: These men were from POW camp Stalag 2B which had originally been at Hammerstein. They were all sick and had left their column to be taken to a hospital. On arrival at the hospital, they were denied admission and continued the march with little or no rations. These men appeared on the verge of exhaustion.

Two had obvious fevers with severe cough that was probably pneumonia or tuberculosis. About 20 of these men were Americans. On had a foreign uniform and I thought he was Italian. There was a tall British sergeant with them. One of the men carried a small wooden chest with the name "Joe McDaniels" or "Joe McWilliams" on it. He told me that he had been acting Chaplain at Stalag 2B. Another man was a tall, slender fellow from Schenectady, New York. (After I was liberated, I met an ex-prisoner from Stalag 2B who thought this fellow was J. Luckhurst of 864 Stanleyh, Schenectady, New York.)

This fellow said he was suffering from recurrent malaria. These men were so weak they could scarcely stand. The German sergeant in charge of our section at the time recognized their plight and got a Wehrmacht truck to take them to our next stop. We received no rations that night but did get a small issue of hot water. The next day these men were placed on wagons and stayed with us. They again received no rations and again were sheltered in crowded barns.

On 18 February 1945, I personally protested to Capt. Weinert about these men, although he had known about previously. I pointed out that these men were exhausted and might soon die. I requested rations, rest, and hospitalization for them. Capt. Weinert said they were not his responsibility, inasmuch as they were not originally from Stalag Luft IV. I objected to this and stated that these men were now in our column and that he was responsible for their lives and health. He then agreed to leave these men behind.

The next day, Capt. Weinert told me these men had been transferred to another command. I never saw the men again, but I heard a rumor that one of them died.

Q. Do you have anything further to add?
A. No.

Leslie Caplan, M.D.

State of Minnesota
County of Hennepin

I, Dr. Leslie Caplan, of lawful age, being sworn under oath, state that I have read the foregoing transcription of my interrogation and all answers contained therein are true to the best of my knowledge and belief.

(signed)
Leslie Caplan, M.D.

Subscribed and sworn to before me, this 5th day of January 1948.
(signed)
William C. Hoffmann
Lt. Colonel, AGD
Summary Court

CERTIFICATE

I, William C. Hoffmann, Lt. Col. Certify that Dr. Leslie Caplan personally appeared before me on December 31, 1947 and testified concerning war crimes; and that the foregoing is an accurate transcription of the answers given by him to the several questions set forth.

(signed)
Lt. Col. William C. Hoffmann

Place: Minneapolis, Minn.
Date: 31 December 1947

PART III:
STORY OF A SURVIVOR

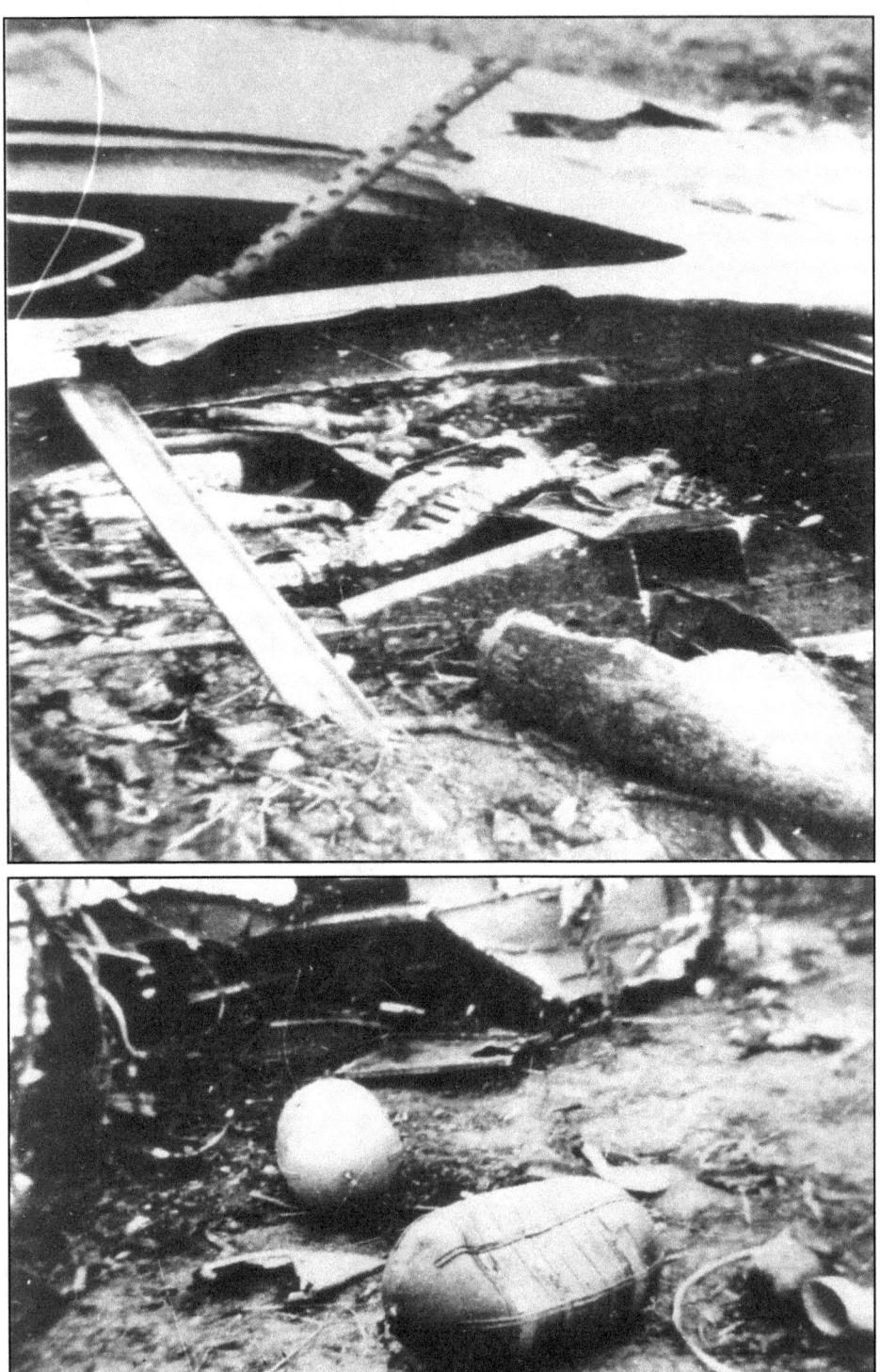

July 2, 1944: Damaged remains of the B-24 on which crew member Fred Meisel was shot down and captured as volunteer replacement.

FRED MEISEL

In July 1944, while in Royal Hungarian Hospital No. 11, I met Fred Meisel. He told us how Hungarian peasants had spared because of a nickel that he carried on his dog chain. Fred was sent to the same POW camp that I was, but I do not know the Lager to which the Germans assigned him.

Meisel was born of Jewish parents in Minsk, Russia, on November 7, 1905. His parents separated and Fred was left with an uncle while his mother went to medical school in Switzerland. He attended school in Berlin and, in 1917, enlisted in the German Army with his entire school class. Fighting on the western front, these boys were involved in some of the bloodiest battles of World War I. As a machine gunner, Fred was awarded the Iron Cross. After the war, Fred stayed in the German Army. In 1922, he immigrated legally to United States where he worked as a truck driver, bricklayer, and a contractor as a private civilian.

When World War II broke out, he enlisted in the U.S. Army on December 29, 1941. He was placed in chemical warfare, but craving combat, he transferred to the infantry, where he fought the Japanese in the South Pacific. Somehow, Fred was able to transfer again winding up in the Aleutians in the Army Air Corps, as a gunner. He flew sixty missions, including the first bombing raids on Paramushiro and Katoaka Naval Bases.

At age forty-two, Fred was perhaps the oldest active aerial gunner in the 15th Air Force. He vowed to fly one hundred missions and again badgered his way into a B-24 bomber in Italy in the 456th Bomb Group. Flying at every opportunity, he soon got ahead of his regular crew. On July 2, 1944, he flew as a volunteer replacement on another crew, and on his 100th mission, Fred was shot down over Budapest, Hungary.[6] Fred was wounded and the plane set on fire, but he helped two other crew members out of the ship before jumping out himself.

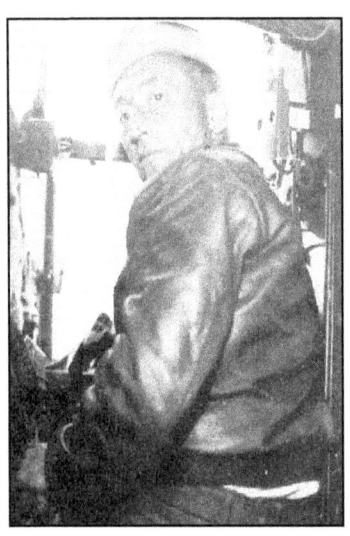

Fred Meisel

Fred upon landing after he bailed out of his burning plane was immediately apprehended by Hungarian peasants carrying scythes, pitchforks and clubs. The peasants wanted to kill Fred since he had a "J" on his dog tag (he was Jewish). He also had an American Indian–head nickel on the chain around his neck that held his dog tag. Fred finally convinced the peasants not to kill him by telling them he was an American Indian. Since his features strongly resembled the Indian on the nickel, they spared his life but not before slicing his little finger lengthwise very badly. He

---

[6] One published account reported that Meisel flew 103 missions as a gunner before he was shot down and taken prisoner.

was placed in a ward with other injured American and British airmen in the Royal Hungarian Hospital No. 11 in Budapest. After spending some time in the hospital, Fred shipped to a POW camp Stalag Luft IV in Germany.

On February 6, 1945, Stalag Luft IV POW camp was evacuated because of the advancing Russian Army. Fred spent the next eighty some days walking through Germany on the same 600-mile march I did, "The Death March of Stalag Luft IV."

Fred was discharged on October 6, 1945, but stayed on the active reserve, where he finally retired from Army Intelligence Service in 1965. He earned virtually every battle star given in World War II. He was probably the oldest gunner flying combat missions and the first to reach 100 missions. His combat career spanned four decades, and he was awarded some of the highest honors.

On June 1, 1946, Fred married Mary Catherine Harwood. They lived in the San Fernando Valley area before moving to Porterville, California. In the 1970's they also lived in Jalisco, Mexico where Fred instructed Judo. He died on February 22, 1987.

PART IV:
BACK TO THE FUTURE

Fifty years after his capture, John L Lenburg holds the spoon the Nazis gave him during his encampment as a POW at Stalag Luft IV. *Photo by Greg Lenburg. Courtesy: John L. Lenburg Collection*

FIFTY YEARS LATER

In 1982, the 460th Bombardment Group Association came into being by holding a reunion in Orlando, Florida. My wife Catherine and I attended this gathering of former members of the 460th. Erf and other crew members also attended. Some of us had not seen each other in 40 years. After this reunion, it became a yearly event Catherine and I usually attended.

On May 16, 1994, some fifty years later, seventeen of us former 460th members, some with their wives, returned to Italy for a tour. We toured Lake Maggiore, Milan, Venice, Florence, Del Gargano, Rome, Bari, Spinazzola, and Ligano, Switzerland. The highlight of the trip was returning to the site of our old air base on May 25 at the Lorusso farm near Spinazzola. Standing there looking out over the peaceful wheat field with tassels swaying in the breeze, it was had to

Top: Me, Jack Nagle, Erf and Mike Brown, 1987. *Bottom:* With Alan Barrowcliff and Mike Brown in front of the All-American B-24J at our 1990 reunion. *Courtesy: John L. Lenburg Collection*

Top: Sparky Bohnstedt, Dan Whistler (Champlin's crew), Betty Bohnstedt, Jerry Conlon (Evans's crew), Lydia Erfedt (Erf's widow), Mike Brown and me at the present of plaques at the Albuquerque reunion. *Bottom:* Fifty years later, me and Anna Lorusso. *Courtesy: John L. Lenburg Collection*

Top: Nandi, me, and my wife Catherine. *Bottom:* Jerry Conlon, Mike Brown, Nandi, Alan Barrowcliff, and me. *Courtesy: John L. Lenburg Collection*

imagine that this site was home for over 2,000 men and sixty-four-engine aircraft fifty years ago. For me it was a very emotional experience, since it had been almost fifty years to the day that I left here on the June 30th Blechhammer mission.

During the trip, Sparky Bohnstedt, the 460th Bomb Group's historian, kept asking many questions concerning the June 30, 1944 mission. He and his wife Betty were putting together a 460th Bomb Group history book. We also visited the building in Spinazzola where the 55th Bomb Wing Headquarters was located. It is now being used as an elementary school. We were mobbed by the schoolchildren as we entered the building waving little American flags. This was another highlight of the trip.

Framed mementos from my Army Air Force days. *Photo by Greg Lenburg*

Miss Fortune

In April 1994, Sparky received a letter from a young college student Nándor Mohos, nicknamed Nandi, requesting information on two aircraft that were lost over Hungary on June 30, 1944. Sparky informed him that four aircraft were lost, not two. As time passed, this enthusiastic young man and I exchanged much information. He became very dedicated to the 460th Bomb Group.

After Nandi graduated from college, he served an obligatory tour of duty with the Hungarian Army as a first lieutenant. While serving with the Hungarian Army, he wrote an article about the 460th that was

An excavation crew recovers the engine to "Miss Fortune" in the Lake Balaton area where it was shot down in June 1944. *Courtesy: Nándor Mohos*

published in the *Hungarian Soldier*, a publication for the Hungarian Army. Through the cooperative efforts of Sparky and Nandi, the crash sites of our plane, Evans' plane, and Champlin's plane were found. Nandi also found a series of 35mm color slides taken from St. Gyorgy Hill, south of Tapolca, on June 30, 1944 of the air battle taking place over Lake Balaton (images which appear in Ch. 21, "Going Down.")

At our annual reunion dinner at Albuquerque, New Mexico on September 23, 1995, along with other survivors of the ill-fated June 30, 1944 mission to Blechhammer, I was presented with a plaque containing an actual fragment of our aircraft, "Miss Fortune." Since meeting Nandi at our reunion in Huntsville, Alabama in 1995, we have corresponded with each other.

The Propeller

It has been proposed to use the recovered piece of propeller from our plane as part of a monument dedicated to the 460th airmen killed over Hungary in World War II. It would be located in the Lake Balaton area of Hungary. The propeller cannot leave the country of Hungary since it is now considered their property.

Left: The recovered propeller of "Miss Fortune." *Right:* Nandi with the propeller on display in Hungary. *Courtesy: Nándor Mohos*

Martin Troy

Martin Troy

In 1987, Alan Barrowcliff sought information regarding where the remains of Troy, Waits and Wheeler were interned. He learned that Wheeler was buried a military cemetery in Lorraine, France. Waits remains were sent home to Atlanta and Troy's in Florence, Italy. At Troy's site there was only a marker.

The finding of Troy's remain was chronicled in the November/December 1991 issue of *Air Progress Warbirds International* magazine in an article entitled, "Liberation of a Liberator," by Mihaly T. Lenart. Lenart was making a documentary TV film about World War II aircraft that crashed in Hungary and what happened to them. In the article he talked about recovering a part of a propeller at the crash site of our plane, "Miss Fortune." He said people told him of bones being found in the area. Since there was only a marker with Troy's name at the military cemetery in Florence, Italy, we theorized his remains could be still at the crash site.

Nandi and I felt the felt the same way, but he researched the death records in the County of Veszpreme and reports of the day's events to the Lord Lieutenant of the County of Zala. In these reports, only Waits and Wheeler's name were mentioned. With this information, I enlisted the help of Congressman Peter Visclosky's office. He had Troy's records pulled from the Washington National Records Center, in Maryland. These records told us that a search was made of the death and cemetery records in the area. A search of the crash site area was also made but because of the high lake water level and the political climate only a cursory search was done and nothing was found.

Next, we went to the Mortuary Affairs and Casualty Division of the U.S. Army in Alexandria, Virginia. I eventually received a letter from Lt. Col. Robert Steward indicating they would not send a recovery team to the area without better-substantiated evidence of Troy's remains being there.[7]

The All-American B-24J

At our 1990 460th reunion in Savannah arrangements had been made to have a restored B-24J flown in for our reunion. This aircraft had been refurbished and made flyable by The Collings Foundation, a non-profit organization in Boston, Massachusetts which since 1989 has restored and exhibited aircraft from World War II as part of its "Wings of Freedom Tour." It had taken a number of years to complete this project.

---

[7] *In August 2007, the remains of Martin Troy were found among the wreck of a B-24H Liberator bomber in the village of Nemesvita, about 110 miles southwest of the capital Budapest. After confirming the identity, his remains were returned to the United States for internment on November 20, 2008 at Arlington National Cemetery.*

Conseil Régional de Basse Normandie

Le Directeur de Cabinet

Mr JOHN L. LENBURG
PORTAGE
IN 46368-9650
USA

Caen, 1st October 1997

Dear Sir,

I have the pleasure and honour to inform you that President René GARREC has decided to award you the Jubilee of Liberty medal. In this exceptional case we are sending you the Jubilee of Liberty medal, offered by the regional council of Lower Normandy. The certificate is not available any more.

Sincerely Yours,

Philippe FORIN

Letter from the regional council of Lower Normandy in France honoring me with the Jubilee of Liberty Medal in 1997. *Courtesy: John L. Lenburg Collection*

Top: With my nephew Mark Prusinski in front of the All-America B-24J we flew. *Bottom:* In my home office surrounded by remnants of my POW past. *Courtesy: John L. Lenburg Collection*

We were given the opportunity to go all through the aircraft at the reunion. Each year, this aircraft tours the U.S. and comes to Indiana and I have been able to see the plane.

On August 11, 1998, I was given the opportunity to take an hour's flight in the aircraft. My nephew Mark Prusinski made the arrangements. The morning we got up early, at 4:30 A.M., and to Fort Wayne, Indiana. So 53 years and 1-½ months after my last flight, June 30, 1944, I was able to fly in a B-24 again. It brought back a lot of memories, some good and some bad. While in flight, they allowed us go all over the plane, except into any of the turrets.

AFTERWORD

In 1998, John L. Lenburg wrote and published, *Kriegsgefangenen #6410 (Prisoner of War)*, which detailed his experiences as prisoner of war and paid tribute to his fellow crew members of the 460th Bomb Group, 760th Bomb Squadron, 15th Air Force.

Following its publication, he appeared at local schools and civic organizations to speak about the great sacrifice made by his generation to preserve freedom.

John L. Lenburg was a patriot to the end. Sadly, after a life dedicated to his country and his fellow man, he died of heart failure on November 4, 2000. His last wish was that this book would be shared so future generations would benefit from his remarkable experiences.

"Diamond Lil'," built in 1941 and the twenty-fifth Liberator produced by Consolidated Aircraft Corporation, is the world's oldest continuously flying aircraft. Here, it is pictured in flight during the Air Force Association's "Gathering of Eagles" convention in Las Vegas, Nevada in 1986.

PHOTO GALLERY

The following is a gallery of B-24 photographs and wartime memorabilia from special collections.

Vintage 1940s postcard supporting the war effort and American servicemen abroad during the Second World War.

An original Coca-Cola print ad featuring the Consolidated B-24 Liberator bomber. *Courtesy: The National Air and Space Museum Poster Collection*

The "Joisey Bounce" leads a trio of B-24D Liberators from the 93rd Bomb Group in formation. *Courtesy: World War Photos*

A U.S. Army Air Corps B-24 Liberator crosses over shark-nosed Curtiss P-40 Warhawk fighter planes to land. *Courtesy: AllPoster.com*

Also from Moonwater Press

SCARED TO DEATH:
A Lori Matrix Hollywood Mystery

Twentysomething ex-swimsuit model-turned television news anchor Lori Matrix wants to prove her producers wrong: She can do hard news. The tall, blonde, blue-eyed journalist, hired to boost her station's ratings, gets her chance when legendary actor Rex Hampton turns up dead at the first gathering of six Hollywood movie legends during the screening of his favorite movie. Hampton becomes the first in a series of macabre murders. Three more movie giants from the group drop dead, under mysterious circumstances, at successive film screenings. Something on the screen, something in those movies, made their hearts stop. An unexplained phenomenon? A ghost? What? Believing all four deaths are related, Matrix proves it, as she unmasks the perpetrator behind the evil in this original Hollywood whodunit.

PEEKABOO:
The Story of Veronica Lake

In the 1940s, Veronica Lake stirred a nation of filmgoers with memorable film performances. The sultry blond bombshell rocketed to stardom only for her life and career to spiral out of control. By age twenty-nine, she was washed up, broke, and destitute. Her fall from grace started with her dominating mother, who saw her insecure daughter as her meal ticket, only to end tragically after Veronica's years of struggling with paranoid schizophrenia and alcoholism. This fully authorized, revised and expanded biography, featuring interviews with Lake's mother, friends, co-workers, and family members and 150 illustrations, tells the complete, unvarnished truth of the rise and fall of one of the silver screen's most beloved stars.

www.ingramcontent.com/pod-product-compliance
Lightning Source LLC
Chambersburg PA
CBHW060109170426
43198CB00010B/825